Morphology and Dynamics of Crystal Surfaces in Complex Molecular Systems

MATERIALS RESEARCH SOCIETY
SYMPOSIUM PROCEEDINGS VOLUME 620

Morphology and Dynamics of Crystal Surfaces in Complex Molecular Systems

Symposium held April 23–27, 2000, San Francisco, California, U.S.A.

EDITORS:

Jim De Yoreo
Lawrence Livermore National Laboratory
Livermore, California, U.S.A.

William Casey
University of California at Davis
Davis, California, U.S.A.

Alexander Malkin
University of California at Irvine
Irvine, California, U.S.A.

Elias Vlieg
University of Nijmegen
Nijmegen, The Netherlands

Michael Ward
University of Minnesota
Minneapolis, Minnesota, U.S.A.

Materials Research Society
Warrendale, Pennsylvania

Single article reprints from this publication are available through
University Microfilms Inc., 300 North Zeeb Road, Ann Arbor, Michigan 48106

CODEN: MRSPDH

Published by:

Materials Research Society
506 Keystone Drive
Warrendale, PA 15086
Telephone (724) 779-3003
Fax (724) 779-8313
Web site: http://www.mrs.org/

Library of Congress Cataloging-in-Publication Data

Morphology and dynamics of crystal surfaces in complex molecular systems : symposium held
    April 23–27, 2000, San Francisco, California, U.S.A. / editors, Jim De Yoreo, William Casey,
    Alexander Malkin, Elias Vlieg, Michael Ward
        p.cm.—(Materials Research Society symposium proceedings,
        ISSN 0272-9172 ; v. 620)
        Includes bibliographical references and indexes.
        ISBN 1-55899-528-5
        1. De Yoreo, Jim II. Casey, William III. Malkin, Alexander IV. Vlieg, Elias V. Ward,
    Michael VI. Materials Research Society symposium proceedings ; v. 620
2001

Manufactured in the United States of America

CONTENTS

*Invited Paper

BIOGENIC AND BIOMIMETIC SYSTEMS

GROWTH AND MORPHOLOGY AT THE OXIDE SOLUTION INTERFACE

*Invited Paper

POSTER SESSION:
MORPHOLOGY AND DYNAMICS
OF CRYSTAL SURFACES IN
COMPLEX MOLECULAR SYSTEMS

MACROMOLECULES

INORGANIC SYSTEMS II—
IMPURITIES AND DEFECTS

*Invited Paper

*Invited Paper

PREFACE

This volume contains selected papers from the first MRS symposium on "Morphology and Dynamics of Crystal Surfaces in Complex Molecular Systems," held April 23–27 at the 2000 MRS Spring Meeting in San Francisco, California. This is a topic of central importance in fields as diverse as geochemistry, crystal growth, structural biology, corrosion science, pharmaceutical production, and food science. In all of these areas, understanding surface morphology and dynamics requires consideration of multiple chemical species, molecular anisotropy, impurities, and the interface between fluid and solid phases. This volume brings together papers from each of these fields that explore common themes in the growth and dissolution of inorganic, organic, and macromolecular crystals and films produced both through natural and synthetic processes. The focus of these papers is on the physical and structural studies of these processes.

Many of the contributions to this volume include the results of atomic force microscopy investigations of surface dynamics. Indeed its emergence as the dominant tool for surface studies of molecular crystals, particularly in solutions where real-time studies are possible, is largely responsible for the recent re-emergence of interest in fundamental research within this field. A number of the papers within this volume reflect the growing sophistication of this research, with the flavor of the investigations showing an evolution from mere "stamp collecting" to detailed studies of step edge fluctuations, impurity step-interactions, and assembly pathways in non-Kossel crystals both during nucleation and step advancement.

A second experimental method which is emerging as a tool of major importance within this field is surface x-ray diffraction, a method for determining the atomic structure of crystal surfaces. While this tool has been used with great success in the field of vapor phase epitaxy of metal and semiconductor films, only in recent years has it been applied to molecular crystals. Most importantly, a number of studies have now demonstrated the power of this tool to investigate the structure of crystal-liquid interfaces on both the solid and liquid side of the interface. This is a critical area of crystal growth science that, until now, has been an arena more for speculation than quantitative science.

Another major development that is reflected in this volume is the advance in molecular-based crystal engineering. The recent foray of organic and synthetic chemists into this field has produced new insights into the role of molecular recognition and anisotropy on crystal structure and growth. Similar insights are emerging from studies of crystallization in biologically relevant systems including biominerals, pharmaceuticals and proteins. Examples from all of these fields are included in this collection of papers.

As this volume demonstrates, investigations of molecular crystallization at the nanoscale have not only found their way into the research repertoire of this diverse set of disciplines, but are continuing to grow in scope and number. Still, many fundamental questions about the growth and dissolution of molecular crystals remain unanswered. As a consequence, we believe this will be just the first in a series of MRS volumes on this topic.

Jim De Yoreo
William Casey
Alexander Malkin
Elias Vlieg
Michael Ward

November 2000

MATERIALS RESEARCH SOCIETY SYMPOSIUM PROCEEDINGS

MATERIALS RESEARCH SOCIETY SYMPOSIUM PROCEEDINGS

Prior Materials Research Society Symposium Proceedings available by contacting Materials Research Society

Inorganic Systems—Surface Morphology and Step Kinetics

Mat. Res. Soc. Symp. Vol. 620 © 2000 Materials Research Society

From the Solid-Fluid Interfacial Structure to Genuine Morphology of Crystals

X.Y. Liu

Department of Physics, National University of Singapore, 10 Kent Ridge Crescent, Singapore 119260.

ABSTRACT

Crystals reveal a large variety of shapes, depending on the chemical composition and the structure of crystals, and the growth conditions. To predict precisely the growth morphology of crystals remains a challenging issue for both academia and industry. In this paper, the principles for predicting the growth morphology of crystals based on the best available theories are highlighted and demonstrated.

INTRODUCTION

The relative distance from the centre of the crystal to the respective crystal faces $\{hkl\}$ h_i determines the shape of a crystal. For a steadily growing crystal, these distances are evidently proportional to the relative growth rates R_{hkl}^{rel} of the crystal faces [1-6]. (See Fig.1.) The implication is that the prediction of the growth morphology is equivalent to the prediction of relative growth rates in different crystallographic orientations.

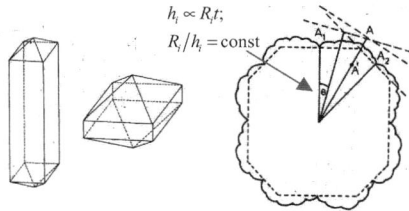

$$h_i \propto R_i t;$$
$$R_i / h_i = \text{const}$$

Figure 1 *The shape of crystal is determined by h_i. $h_i = t R_{hkl}^{rel}$ (t: time). Therefore, $h_i \sim R_{hkl}^{rel}$.*

In order to predict the growth morphology of crystals, some *ad hoc* recipes [1-4] have been published in the past. Taking solely the geometry of crystal lattices into account, Bravais, Friedel, Donnay, Harker put forward the following relation as a recipe for the prediction of growth morphology, [1-4]

$$R_{hkl}^{rel} \propto 1/d_{hkl} \tag{1}$$

where d_{hkl} is the interplanar distance, corrected for the extinction conditions of the space group.

In 1955, Hartman and Perdok published their famous morphological theory, the Hartman-Perdok (or PBC) theory [2-4]. It is assumed by this theory that a growing crystal is bounded by F (or flat) faces [2-4], and the relative growth rate R_{hkl}^{rel} of an F face is directly proportional to its attachment energy E_{hkl}^{att}, as [2-4]

$$R_{hkl}^{rel} \sim E_{hkl}^{att} \tag{2}$$

The attachment energy E_{hkl}^{att}, corresponds to the energy released per structural unit as the crystal slice (hkl) attaches to the crystal surface from infinite distances.

Although both the HP and the BFDH theories have achieved some success for crystals grown from the vapour phase[3], discrepancies between the theoretical and the observed morphology occur frequently when crystals are grown from solutions or the melt (c.f. Figs.4-6.) The major problems for these recipes can be summarized as follows: (a) The growth of crystal faces is controlled by the kinetics, [3,6,7] which is not unambiguously and explicitly taken into account in Eqs. (1) and (2)

(b)The surface kinetics is determined by the structure of interfaces between the crystal and the fluid phase. As indicated by interfacial modelling and experiments [10-15], the solid-fluid interfaces are normally inhomogeneous, and the ordering of fluid molecules at the crystal surface causes the adsorbed fluid units to adapt themselves to the crystal surface in certain conformations and configurations very different those in the bulk phase. Unfortunately, these have not been taken into account in the above theories. Obviously, these inherent drawbacks cause the deviations between the predicted and observed morphology of crystals.

PRINCIPLE FOR THE PREDICTION OF MORPHOLGY OF CRYSTAL

According to theories of crystal growth [3,6,7], the growth of faceted crystal faces is governed in many cases by the screw dislocation mechanism. Based on this mechanism, we arrived [8,9] at the relation between R_{hkl}^{rel} and habit controlling parameters for the case of crystals grown from solutions as

$$R_{hkl}^{\text{rel}} \propto d_{hkl} n_{hkl} X_A^{C_{\ell(hkl)}} \exp(-C_{\ell(hkl)}^* \xi_{hkl} \Delta H^{\text{diss}} / 2n_{hkl} kT) [\xi_{hkl} C_{\ell(hkl)}^*]^{-1}$$

$$\times \exp(-C_{\ell(hkl)}^* \xi_{hkl} \Delta H^{\text{diss}} / n_{hkl} kT). \tag{3}$$

In deriving Eq.(3), relations between the structure of crystal/fluid interfaces and surface kinetics controlling parameters, such as the step free energy, kinetic coefficient of step integration, kink density of steps, were established with the framework of the screw dislocation mechanism. Therefore, the calculation of these kinetic parameters becomes possible. For details, see Refs. [8,9]. In the equation, d_{hkl} is the interplanar distance of the crystallographic orientations $\{hkl\}$, X_A the activity of the solute in the bulk, ΔH^{diss} the molar enthalpy of dissolution, k Boltzmann's constant, T temperature, n_{hkl} a constant which is approximately equal to the coordination number of a structural unit within the crystal slice (hkl). ξ_{hkl} is a crystallographic orientation factor, defined as

$$\xi_{hkl} = 1 - E_{hkl}^{\text{att}} / E^{\text{cr}}, \tag{4}$$

and $C_{\ell(hkl)}^*$ denotes the surface scaling factor, defined as the ratio between the local enthalpy of dissolution at the crystal surface (hkl) $\Delta H_{hkl}^{\text{diss}}$ and ΔH^{diss} [6,8-10],

$$C_{\ell(hkl)}^* = \Delta H_{hkl}^{\text{diss}} / \Delta H^{\text{diss}}. \tag{5}$$

Eq. (3) is a basic equation in the prediction for the growth morphology of crystals associated with screw dislocation controlled growth. In the case where other growth mechanisms become dominant, different equations [5] should therefore be applied, instead of Eq.(3). Note that the factor d_{hkl} is determined by the structure of the crystals, ξ_{hkl} can be calculated based on Eq.(4), n_{hkl} can be obtained from the connected net of $\{hkl\}$ derived from a PBC analysis, X_A and ΔH^{diss} are determined by the growth conditions.

It can be seen from Eq.(3) that $C_{\ell(hkl)}^*$ has a significant impact on R_{hkl}^{rel} for a given crystal surface. Its value depends strongly on the concentration and the composition of solutions, crystallographic orientations and other crystal growth conditions [10-14]. In other words, $C_{\ell(hkl)}^*$ is one of the most important factors capturing the effect of the fluid phase on the growth kinetics and morphology of crystals. Therefore, one of the key steps in the morphological prediction is to calculate $C_{\ell(hkl)}^*$ under given crystal growth conditions.

According to the interfacial model developed recently [6,8-10], $C^*_{\ell(hkl)}$ can be related to the activities of the solute in different regions as follows

$$C^*_{\ell(hkl)} \approx \ln X^{eff}_{A(hkl)}/\ln X_A, \tag{6}$$

where $X^{eff}_{A(hkl)}$ is the activity of adsorbed growth units at the growing crystal surface interface. (See Fig.2.) The occurrence of the surface activity of growth units is caused by the interactions between the crystal surface and adsorbed fluid units. In the bulk phase, molecules can randomly transfer from one conformation and orientation state to another. On average, a crystal surface in a distant away from these molecules will "feel" the "mean field" of these molecules. In other words, molecules of the same type will regarded to be equal from a statistical point of view. However, the situation will be completely different at crystal surfaces. The specific molecular arrangement and (unsaturated) intermolecular interactions at the crystal surface will cause the adsorbed molecules to adopt a certain orientation and conformation. (See Fig.2.) This means that the orientation and conformation of these molecules will be "frozen" or partially frozen to a certain state at the surface. This leads to the reduction of the effective concentration growth units in the surface kinetics (see the caption of Fig.3a.)

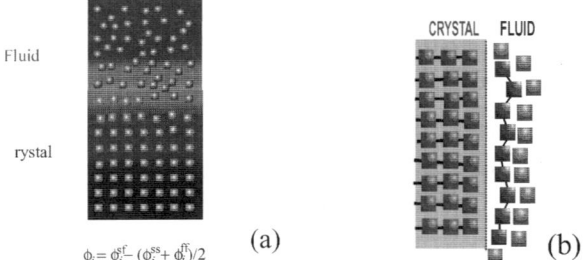

Fluid

rystal

$\phi_i = \phi_i^{cr} - (\phi_i^{ss} + \phi_i^{fl})/2$ (a)

CRYSTAL   FLUID

(b)

Figure 2. *Illustration of the ordering of spherical molecules and chainlike molecules at the crystal surface.*

According to the definition, the surface activity of adsorbed growth units can be calculated by [15-16]

$$X^{eff}_{A(hkl)} = \int_{\omega,q} \varsigma(\omega,q) X_{A(hkl)}(\omega,q) d\omega dq = \delta_{hkl} X_{A(hkl)}, \tag{7}$$

with the activity coefficient,

$$\delta_{(hkl)} = \int_{\omega,q} \varsigma(\omega,q) P_{A(hkl)}(\omega,q) d\omega dq. \tag{8}$$

$$P_{A(hkl)}(\omega,q) = X_{A(hkl)}(\omega,q)/X_{A(hkl)}, \tag{9}$$

and $\quad \varsigma(\omega,q) \approx \exp(\Delta G^*_{kink}/kT)/\{\exp(\Delta G^*_{kink}/kT) + \exp[\Delta G^*(\omega,q)/kT]\}. \tag{10}$

where ω and q denote the orientation and conformation of an adsorbed growth unit, respectively; ΔG^*_{kink} denotes the dessolvation energy at kink sites, and $\Delta G^*(\omega, q)$ is the free energy barrier for the growth unit to obtain the "correct" orientation and conformation in the step integration. We note that $\varsigma(\omega, q)$ denotes the activity coefficient at the given orientation and conformation state. The calculation of the activity of adsorbed molecules is illustrated in Fig.3b.

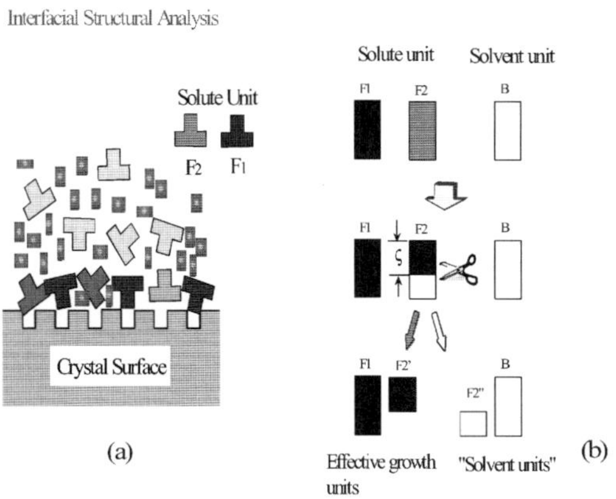

Interfacial Structural Analysis

Solute unit   Solvent unit

Solute Unit

F2   F1

F1   F2   B

F1   F2   B

ς

Crystal Surface

F1   F2'   B

Effective growth units   "Solvent units"

(a)   (b)

Figure 3 (a) The ordering of fluid molecules at the surface leads to the consequence that some adsorbed growth units can easily be incorporated into the crystal structure at kink sites while others show insurmountable free energy barriers to enter kinks of steps due to their "wrong" orientations or conformations. It follows that we have then the surface activity of growth units $X^{\mathrm{eff}}_{A(hkl)}$ smaller than $X_{A(hkl)}$ the concentration of adsorbed growth units. F_1: adsorbed molecule with a right conformation and orientation. F_2: adsorbed molecule with a wrong conformation and orientation. (b) Illustration of the calculation of the activity of adsorbed molecules.

APPLICATIONS

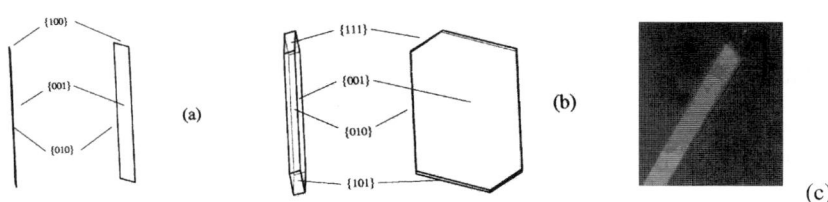

{100}

{001}

{010}

(a)

{111}

{001}

{010}

{101}

(b)

(c)

Figure 4 The morphology of n-$C_{24}H_{50}$ grown from i-octane solution [15,16]. (a) The morphology predicted based on the model given in this paper. (b) The morphology predicted according to the HP theory; (c) The observed morphology.

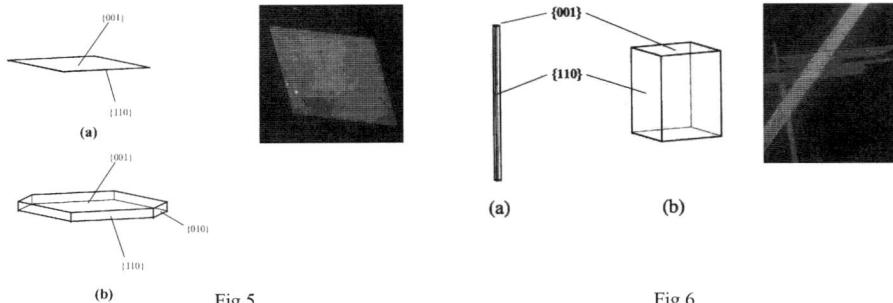

Fig.5   Fig.6

Figure 5 The morphology of n-C$_{21}$H$_{44}$ grown from n-hexane solution. (a) The morphology predicted based on the model given in this paper. (b) The morphology predicted according to the HP theory; (c) The observed morphology [15,16].

Figure 6 The morphology of urea crystal grown from aqueous solution. (a) The morphology predicted based on the model given in this paper. (b) The morphology predicted according to HP theory; (c) The observed morphology [8,9].

Conclusion: our approach gives much precise predictions than the HP theory.

REFERENCES

1. P. Bennema, and J.P. van der Eerden, in *Morphology of Crystals, Part A*, edited by I. Sunagawa (Terra, Sci., Tokyo, 1987) p.1

2. P. Bennema, in *Handbook on Cryst. Growth*, edited by D.T.J. Hurle, (North-Holland, Amsterdam, 1993) p.477.

3. A.A. Chernov, *Modern Crystallography III-Crystal Growth*, (Springer-Verlag, Berlin, 1984) p.1.

4. P. Hartman, in *Morphology of Crystals,* edited by I. Sunagawa, (Terra, Tokyo, 1987), p.269.

5. X.Y. Liu, and P. Bennema, *Phys. Rev.* **B49** (1994) 765; **B53** (1996) 2314

6. X.Y. Liu, and P. Bennema, in *Current Topics in Crystal Growth Research*, **2** (1995) 451; X.Y. Liu, *J. Chem. Phys.* **102**, 1373 (1995).

7. W.K. Burton, N. Cabrera and F.C. Frank, *Phil. Trans. R. Soc.* London **A243** (1952) 299; P. Bennema, and J. Gilmer, in *Crystals Growth: An Introduction,* edited by P. Hartman, (North- Holland, Amsterdam, 1973) p.263.

8. X.Y. Liu, E.S Boek, W.J. Briels and P. Bennema, *Nature* **374** (1995) 342.

9. X.Y. Liu, E.S Boek, W.J. Briels and P. Bennema, *J. Chem. Phys.* **103**, (1995).

10. X.Y. Liu, and P. Bennema, *J. Chem. Phys.* **98** (1993) 5863.

11. F.F. Abraham, and J.Q. Broughton, *Phys. Rev. Lett.* **56** (1986) 734 .

12. A.D. Haymet, and D.W. Oxtoby, *J. Chem. Phys.* **74** (1981) 2559.

13. J.Q. Broughton, and G.H. Gilmer, *J. Chem. Phys.* **84** (1986) 5741; **84** (1986) 5749; **84** (1986) 5759.

14. M. Da Silva Couto, X.Y. Liu, H. Meekes, and P. Bennema, *J. Appl. Phys* **75** (1994) 627.

15. X.Y. Liu and P. Bennema, in *Current Topics in Crystal Growth Research 2*, 451 (1995).

16. X.Y. Liu, *Phy. Rev.* ***B60***, 2810 (1999).

Fluctuations of Step Positions at KDP Crystal Faces

Leonid N. Rashkovich, Oleg A. Shustin, Tamara G.Chernevich
Physics Department, Moscow State University, Moscow, 119899 Russia.

ABSTRACT

Applying atomic force microscopy in single string scanning mode, dependence of fluctuations of step shifts on KDP crystal faces on time was determined during crystallization from solution. It was shown that fluctuation amplitude grows proportionally to $t^{1/4}$. On prism faces fluctuations at dissolving were bigger than at growth and bigger than on bipyramid faces. Amplitude of fluctuations was not dependent on distance between steps. Interpretation of experimental results and computation of elementary parameters of crystallization was done based on theory of V.V.Voronkov.

INTRODUCTION

Velocity fluctuations of elementary growth layers, related to fluctuations of the number of building units attached from the media during crystallization of inorganic compounds from solutions were not investigated as of yet. Similar research was done only at high temperatures under vacuum on surfaces of metals and semiconductors [1-4]. Theoretical interpretation of these works was based on assumption that building units are delivered to locations of growth due to surface diffusion [5,6]. In condensed media crystal growth may differ substantially from growth in gas phase as surface diffusion of building units to step risers, where they are incorporated into the crystals, will not necessarily play any significant role.

Investigation of fluctuations in processes of crystallization seems to be important at least in three aspects.

Firstly, comparison of experimental results to microscopic theoretical model allows to make a good judgement of its adequacy to observed process. If the model is selected, it is possible to make an attempt to determine its parameters. In particular, such fundamental parameters like kink density, ρ, frequencies of building units attachment to kinks, ω^+, (and detachment from them - ω^-), linear free energy of steps, α_l, in most cases may be computed only based on fluctuation research. The problem is that direct observation of kinks and building units incorporating into them is only possible when kinks are not numerous, building units are big enough and attachment frequencies are comparatively low. In solutions it is possible only for crystals of macromolecular compounds [7-9].

Secondly, fluctuations may lead to coagulation of elementary steps into bigger ones - macrosteps. Fluctuating, a step for some time retains one position periodically retreating from it. This time τ_f may be determined if dependence of fluctuation amplitude on time (for example, $w=(\chi t)^\beta$, where w - average squared shift of step segment, χ and β are constants) and step velocity v are known. Then τ_f may be found from equation $(\chi \tau_f)^\beta \approx v\tau_f$. If within time τ_f fluctuation shift reaches half of the distance between steps, then macrosteps will be formed.

And thirdly, fluctuations are important from the view of even distribution of impurities encaptured by crystal. Let us assume time of impurity desorption from the upper surface layer of crystal is τ_d, then if $\tau_f > \tau_d$ involvement of impurities will proceed in equilibrium conditions, and impurity will be evenly spread within the volume of crystal. Otherwise the crystal will not be uniform.

Measurements required for analysis of fluctuations assume determination of two dependencies: dependence of average squared position shift of a certain step point on time and dependence of average squared difference of positions of two points of step on distance between them at the same moment of time. Unfortunately AFM is only suited for measurements of the first type, as it has only one measuring tip.

This work is devoted to research of step position fluctuations on KDP crystal faces in solutions close to saturation point. We used atomic force microscope of Nanoscope III type with standard liquid cell from Digital Instruments. Experiments were conducted at 25- 29°C.

EXPERIMENT

Figure 1 shows an image of steps on bipyramid faces. Under high magnification it is visible that they are very cut: it is not possible to distinguish between separate kinks and flat areas among them. At higher resolution, when molecular structure of surface may be seen (figure 2), steps may not be distinguished due to their cuttiness - in presence of a step the whole image looks vague. It is very contrasting to surface images of protein crystals, where under molecular resolution separate kinks on steps are seen distinctively.

Figure 3 shows a series of images fixing position change of a small segment of one step at prism face with time. Slow movement of scanner was switched off, therefore scanning was done along one line: on the chart horizontal axis scales the distance from selected step segment to the left edge of snapshot, vertical axis - time. Alongside with fluctuation, movement of step as a whole is visible. It happened because we could not manage to fix temperature in the cell with required precision. In order to keep the step within the view we had to change intentionally the temperature slightly and thus switch from growth to dissolving and backward.

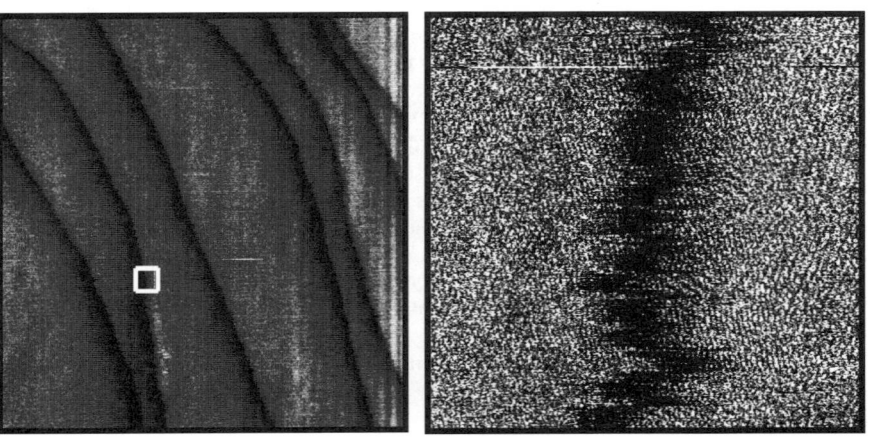

$800 \times 800 \text{ nm}^2$

2

Figure 1*. Steps on bipyramid face. The right chart shows the area, marked with a light square on the left image.*

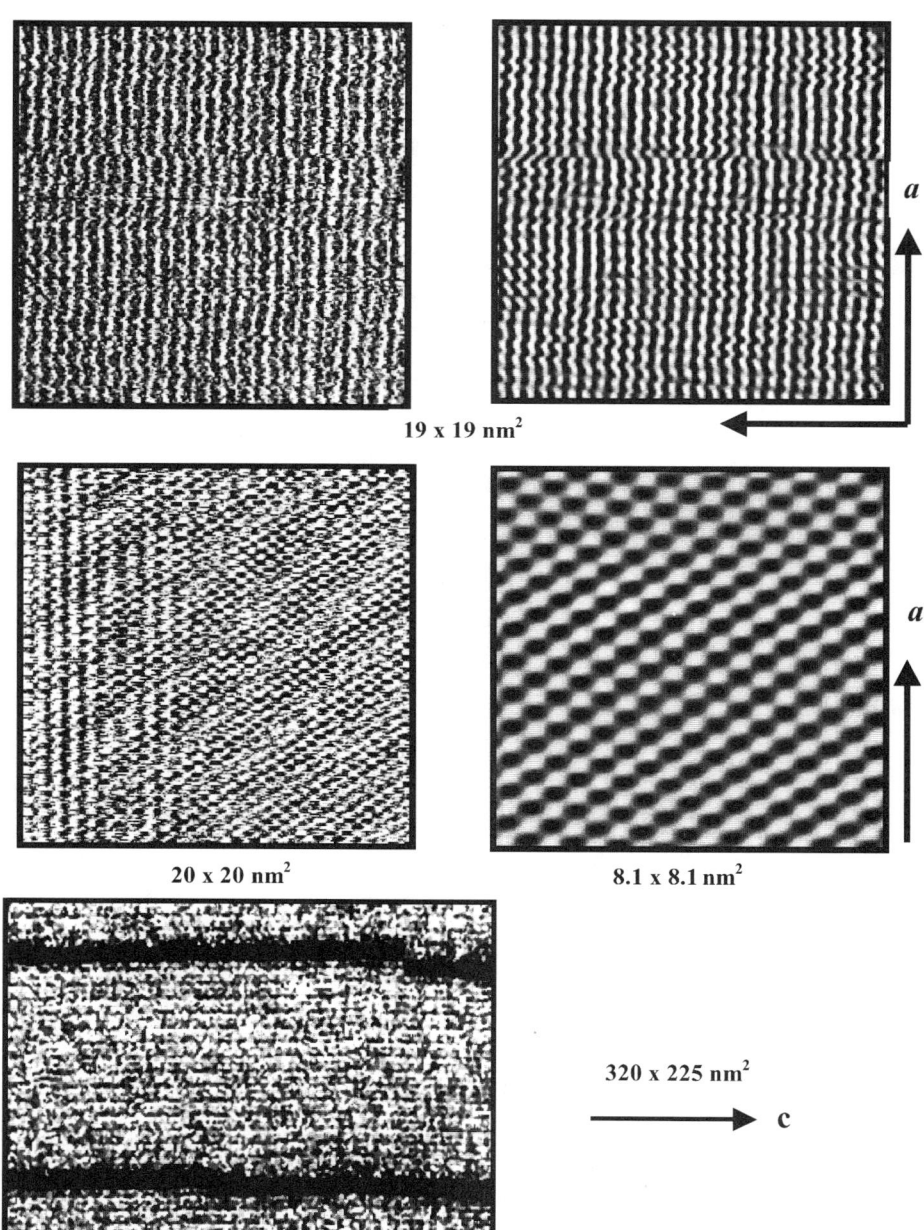

Figure 2. *Surface images with molecular resolution. 1 - prism face; 2 - bipyramid face. On the left - original, on the right – after Fourier transformation, 3 - face (010) of lysozyme crystal [9].*

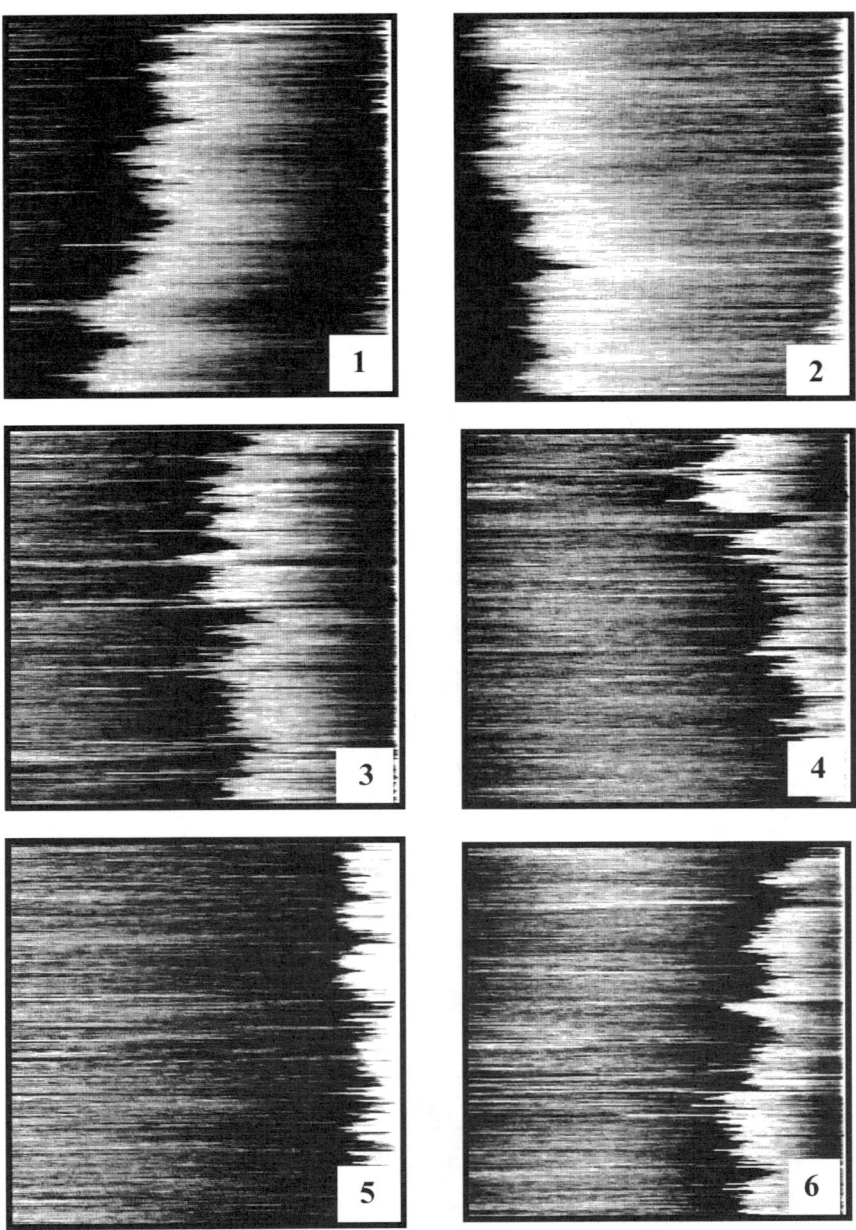

Figure 3. *Position change of step segment on prism face with time. 150 nm x 50.2 s, 512 scans, 10.2 Hz. Light areas - upper terrace, dark areas - lower, therefore step movements to the left correspond to growth (shots 1,2,6), to the right - dissolving (shots 3,4,5). Time of obtaining images, s: 1-0; 2-50; 3-253; 4-304; 5-354; 6-456.*

Figure 4 demonstrates dependence of selected step segment coordinate on time. It was built based of data of figure 3 processed by software program Femtoscan [10], allowing to draw relief section by a single scanning. Judging by inclination of segments of this curve, step velocity did not exceed 0.5 nm/s.

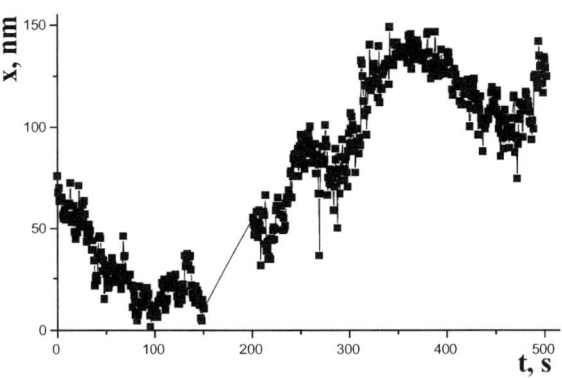

Figure 4. Dependence of x coordinate of selected step segment on time (left edge of snapshots of figure3 was used as beginning of coordinates). Fluctuations are seen on the background of changes, related to drift, growth and dissolving of step.

Bases on data of figure 4 separately for growth and dissolving we built dependencies $<[x(t+\Delta t)-x(t)]^2>$ on Δt, describing growth of fluctuation amplitude within time Δt. Averaging was done for all values of t at Δt = const. If $\Delta t < 4s$ step movement and instrument drift have practically no effect on the measured fluctuation amplitude. These autocorrelation functions are presented in figure 5 along with similarly computed dependence for steps on bipyramid face. In the last case, no difference in fluctuation amplitude at growth and dissolving was observed. Please note that for all Δt distribution function for $<[x(t+\Delta t)-x(t)]>$ was of Gauss type.

Figure 6 presents a series of images, similar to figure 3, of several steps on bipyramid face. Distances between steps are different, which is interesting in view of determining the effect of distance between steps on fluctuation amplitude. Dependence of coordinates of the three steps shown in figure 6 (3, 4 and 5-th counting from the left) on time is charted in figure 7. Figure 8 shows changes of distances between steps 4 and 3, and 5 and 4 with time. On the lower curve of figure 8 segments with duration of about 20-30 s may be distinguished, when distance between

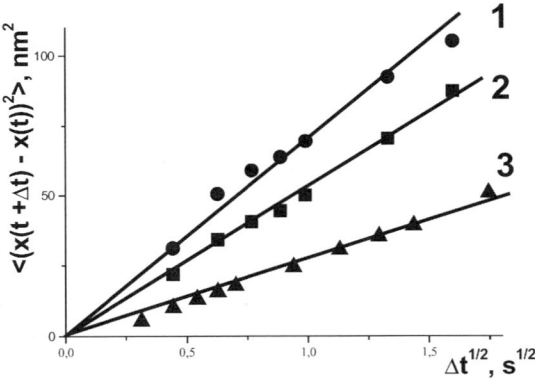

Figure 5. *Dependence of average squared fluctuation amplitude on* $\Delta t^{1/2}$. *Step on prism face: 1 - dissolving, 2 - growth. Steps on bipyramid face - 3.*

steps 5 and 4 is approximately 20, 35, 45 and 60 nm. For all the three steps at small Δt autocorrelation functions practically coincide (figure 9). No correlation between fluctuation amplitude and distance between steps was found.

DISCUSSION

Lesser fluctuations on prism face at growth compared to dissolving (figure 5) may be explained by presence of adsorbed impurity stoppers on the surface. Usually particles of impurities hinder step only at growth and do not affect its movement at dissolving. This statement is well illustrated by figure 10, showing a step by-passing a particle positioned on the surface. It seems to be obvious that if ends of a step are fixed by stoppers, fluctuations of the middle of the step shall be the bigger, the longer the step is.

Bipyramid face has substantially less absorbed impurities, therefore at growth and dissolving fluctuations are equal. Smaller fluctuations on bipyramid face compared to prism face we explain by shorter number of kinks at steps: it is known that steps on these faces are significantly less rounded [11].

The theory of fluctuations at growth from melt and solution was developed by V.V. Voronkov quarter of a century ago [12]. The theory uses three independent phenomenological parameters: β_{st} - kinetic coefficient of step, α_l and $\alpha_e = \alpha_l + d^2\alpha_l/d\varphi^2$ – stability coefficient, describing increase of free energy at small deviation of step from straight-lined shape (here φ - inclination angle of step to tightly packed row of building units, derivative is taken at $\varphi = 0$). At micro level these parameters may be derived through elementary characteristics of crystallization process: ρ and ω^-:

$$\beta_{st} = h^2\omega^-\rho \,, \qquad (1)$$

$$\alpha_e = k_B T/h^2\rho \,, \qquad (2)$$

where h - distance between rows of building units forming a step, k_B - Boltzmann's constant; T - temperature.

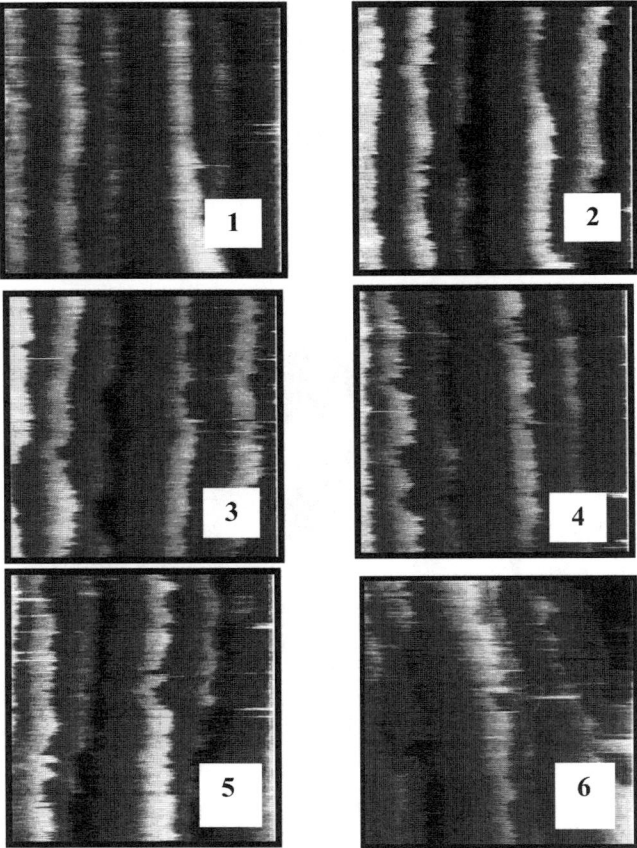

Figure 6. *Image of a segment of several steps of bipyramid face. 250 nm x 25.2 s, 512 scans, 20.3 Hz. 1,2,3 - growth; 4,5,6 - dissolving. Time of obtaining images, s: 1-0; 2-25; 3-50; 4-76; 5-101; 6-126.*

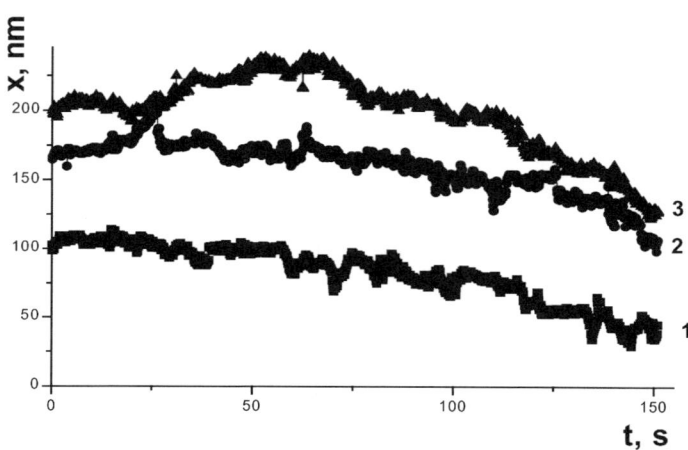

Figure 7. *Change of coordinates with time for segments of three steps shown in figure 6. 1 - step no.3, 2 - step no.4; 3 - step no.5.*

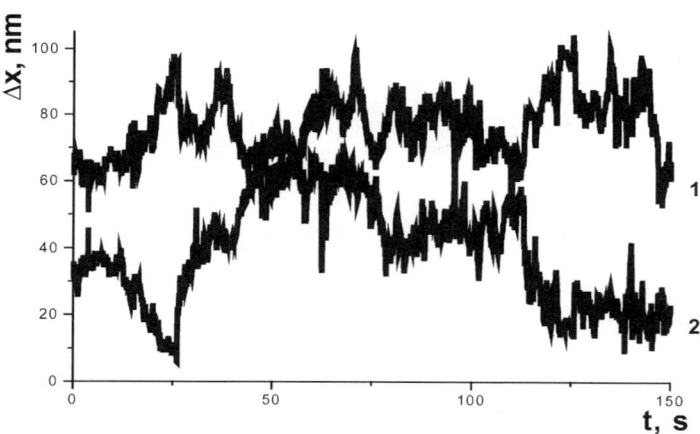

Figure 8. *Development of distance between steps with time (see figures 6 and 7). 1 - between steps 4 and 3; 2 - between steps 5 and 4.*

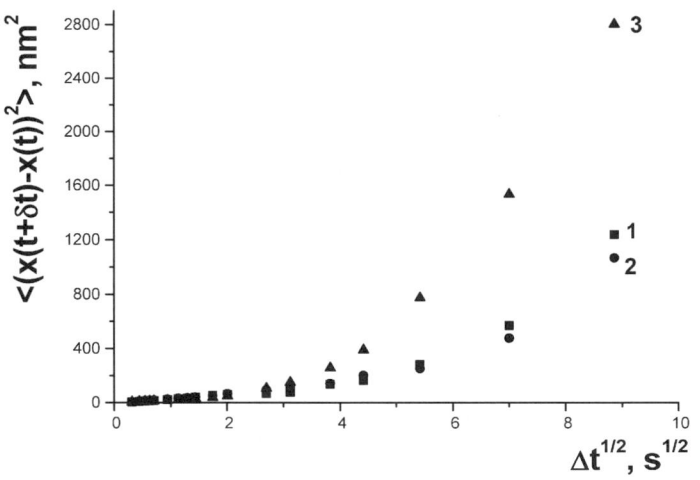

Figure 9. *Autocorrelation functions for curves of figure 7.*

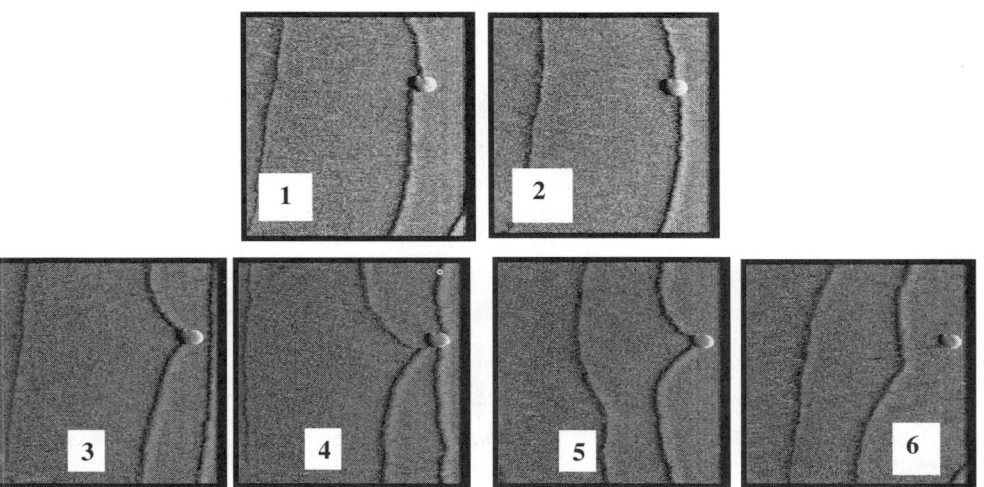

Figure 10. *Step bypassing obstacles on prism face. Size of images 1x1 μm². 1,2 - dissolving, steps move to the right; 3 - 6 - growth steps move left. Time of obtaining images, s: 1 - 0; 2 - 100; 3 - 250; 4 - 300; 5 - 400; 6 - 450.*

Note that attachment frequency of building units to steps (ω^-) at equilibrium state is equal to frequency of detachment, while supersaturation s = $\omega^+/\omega^- - 1$, and step velocity

$$v = \beta_{st}\Omega C_e s, \tag{3}$$

where Ω - volume of molecule in crystal, C_e - equilibrium volumetric concentration of molecules in solution. For KDP at 27°C $\Omega C_e \approx 0.1$.

Following Voronkov, average value of squared fluctuations depends on time in accordance to

$$<w^2> = (\chi t)^{1/2}, \tag{4}$$

$$\chi = 2\beta k_B T/\pi q \alpha_e, \tag{5}$$

where q - surface density of building units in elementary layer, laid during step movement. Considering (1) and (2) formula (5) may be re-written differently:

$$\chi = 2h^4\omega^-\rho^2/\pi q. \tag{6}$$

For tetragonal KDP crystals lattice parameters are $a = 7.45 \cdot 10^{-8}$ cm, $c = 6.97 \cdot 10^{-8}$ cm, consequently for prism face $q = (ac)^{-1} = 19.3 \cdot 10^{14}$ cm^{-2} and for steps moving towards axis c, h = c. For these steps using data of figure 5 we determined $\chi \approx 5 \cdot 10^{-25}$ cm^4/s, thus for T$\approx$300 K ($k_B T \approx$ $410 \cdot 10^{-16}$ erg) out of (5) and (6) we get

$$\beta/\alpha_e = 3.7 \cdot 10^4 \text{ cm}^2/\text{s·erg}; \qquad \omega^-(a\rho)^2 \approx 3.6 \cdot 10^5 \text{ s}^{-1}. \tag{7}$$

Based on figure 1 we could assume that degree of step coverage by kinks $a\rho \approx 0.3$. Then $\beta = 0.078$ cm/s, $\alpha_e = 2.1 \cdot 10^{-6}$ erg/cm and $\omega^- = 4 \cdot 10^6$ s^{-1}, free energy of step riser may be roughly estimated as $\alpha_e/a \approx 28$ erg/cm^2. Assuming step velocity to be 0.5 nm/s we find that supersaturation (and undersaturation) was not higher s = v/$\beta\Omega C_e \approx 6 \cdot 10^{-6}$.

All the mentioned values look true enough. So, literature data for free energy of prism face - 33 erg/cm^2 [13], for kinetic coefficient at 31°C (7.8 ± 1.7)$\cdot 10^{-2}$ cm/s [11, p.138].

At the same time from the above formulas we can only define a combination of two unknown variables. Theory [12] allows to determine each of them if not timely but 2D space dependence of fluctuations is known. For this case the following equation shall be true

$$<(x_i - x_k)^2> = |y_k - y_i| k_B T/\alpha_e = |y_k - y_i| h^2\rho. \tag{8}$$

Here average squared difference of shifts of ends of step segment with coordinates y_i and y_k measured at the same moment of time, is proportional to length of the segment. Meanwhile such measurements are not possible as coordinates of two points of step are determined by nanoscope in different moments of time. Therefore it is now actual to develop the theory further aiming to receive an equation of (8) type when coordinates of i and k points are determined in different moments of time. We would also like to point out that as of yet there is no theory considering the presence of impurity stoppers on the surface.

CONCLUSIONS

Investigation of fluctuation shifts of steps during growth of KDP crystals from solution has shown that this way can deliver important results for science and technology. Was made an assessment of elementary parameters of crystallization: attachment and detachment frequencies of building units, kink density, free linear energy of steps. It was also shown that distance between steps does not affect fluctuation amplitude, which is perhaps an evidence of absence of any significant effect of surface diffusion on crystallization kinetics.

ACKNOWLEDGEMENTS

The authors would like to express their gratitude to A.A.Chernov and V.V.Voronkov for their interest to this work and beneficial discussions. The work was partially supported by grants of the Russian Foundation of Fundamental Research no. 00-02-16701 and NATO PST.CLG. no. 975240.

REFERENCES

1. M. Giesen-Seibert, R. Jentjenns, M. Poensgen, H. Ibach, *Phys. Rev. Lett.* **71**, 3521 (1993).
2 L. Kuipers, M.S. Hoogeman, J.W.M. Frenken, *Phys. Rev. Lett.,* **71**, 3517 (1993).
3. L. Kuipers, M.S. Hoogeman, J.W.M. Frenken, H. van Beijeren, *Phys. Rev.*, **B52**, 11387 (1995).
4. K. Sudoh, T. Yoshinobu, H. Iwasaki, E.D. Williams, *Phys. Rev. Lett.*, **80**, 5152 (1998).
5. A. Pimpinelli, J. Villain, D.E. Wolf, J.J. Metois, J.C. Heyraud, *Surf. Sci.*, **295**, 143 (1993).
6. T. Ihle, C. Misbah, O. Pierre-Louis, *Phys. Rev.*, **B58** , 2289 (1998).
7. A.I. Malkin, Y.G. Kuznetsov, T.A. Land, J.J. DeYoreo, A.P. Barba, J. Konnert, *Biophysical Journal,* **72**, 2357 (1997).
8. Yu.G. Kuznetsov, A.I. Malkin, A. McPherson, *J. Crystal Growth*, **196**, 489 (1999).
9. L.N. Rashkovich, N.V. Gvozdev, I.V. Yaminski, *Crystallograpy Reports,* **43**, 696 (1998).
10. A.C. Filonov, I.V.Yaminski, *http:\\spm.genebee.msu.su.*

11. L.N. Rashkovich. KDP – family Single Crystals. Adam Hilger: Bristol, Philadelphia and New York (1991).

12. V.V. Voronkov, In: *Growth of Crystals*, **11**, Eds. A.A. Chernov, Kc.S. Bagdasarov, E.I. Givargizov, R.O. Sharkhatunyan, Plenum: New York (1979), p.364; V.V. Voronkov. In: *Crystals. Growth, Properties, and Applications*, **9**, Modern Theory of Crystal Growth I, Eds. A.A.Chernov and H.Muller-Krumbhaar, Springer-Verlag: Berlin. Heidelberg, New York (1983) p.7.

13. T.A. Land, J.J. DeYoreo , T.L. Martin, G.T. Palmor, *Crystallograpy Reports,* **44**, 704 (1999).

Fluid-Mineral Interfaces

Mat. Res. Soc. Symp. Vol. 620 © 2000 Materials Research Society

Atomistic Simulations of the $(10\bar{1}4)$ Surface of Carbonate Minerals

Kate Wright[1], Randall T. Cygan[2], and Ben Slater[1]
[1]Royal Institution of Great Britain,
London, W1X 4BS, U.K.
[2]Geochemistry Department, Sandia National Laboratories,
Albuquerque, NM 87185-0750, U.S.A.

ABSTRACT

Atomistic simulation methods have been used to model the structure of the $(10\bar{1}4)$ surfaces of calcite, dolomite, and magnesite under dry and wet conditions. The potential parameters for the carbonate and water species contain shell terms to model the polarizability of the oxygen atoms. These static calculations show that the surfaces undergo relaxation leading to the rotation and distortion of the carbonate groups with associated movement of cations. The dry surface energies are 0.322, 0.247, and 0.256 Jm^{-2} for calcite, dolomite, and magnesite respectively. The influence of water on the surface structure and energies has been investigated for monolayer coverage. When fully hydrated with a monolayer of water, the surface energy for calcite is reduced indicating a stabilization of the surface with hydration. The extent of carbonate group distortion is greater for the dry surfaces compared to the hydrated surfaces, and for the dry calcite relative to that for dry magnesite.

INTRODUCTION

The morphology and atomistic description of calcite surfaces and those of related carbonate minerals (for example, magnesite and dolomite) is needed for an improved understanding of materials and geochemical processes. Calcite ($CaCO_3$) has been incorporated in numerous industrial applications, is of considerable interest in diagenetic processes and related oil production issues, can selectively entrap contaminant metals from the environment, and more recently has been examined with regard to biomineralization and the engineering of materials for a variety of new applications. Although numerous experimental and spectroscopic studies of calcite morphology and surface structure have been completed [1-3], there is no accurate atomistic model of the calcite surface. Even less is known of the structure of the dolomite and magnesite surfaces. Several recent theoretical studies have developed an atomistic treatment of calcite [4-6] and dolomite [4] which has been used to assess the relative stability of different surfaces and to predict external morphology. The aim of the present study is to develop a detailed model of the common $(10\bar{1}4)$ cleavage surface of calcite, dolomite ($CaMg(CO_3)_2$), and magnesite ($MgCO_3$), and evaluate the differences of surface relaxation for the vacuum surface and that modified by water.

THEORETICAL APPROACH

Our theoretical treatment of the $(10\bar{1}4)$ surface of the carbonate phases is based on an ionic model of atomic interaction that incorporates an electrostatic energy term, a Buckingham

Plan view
of top layer
of surface (L1)

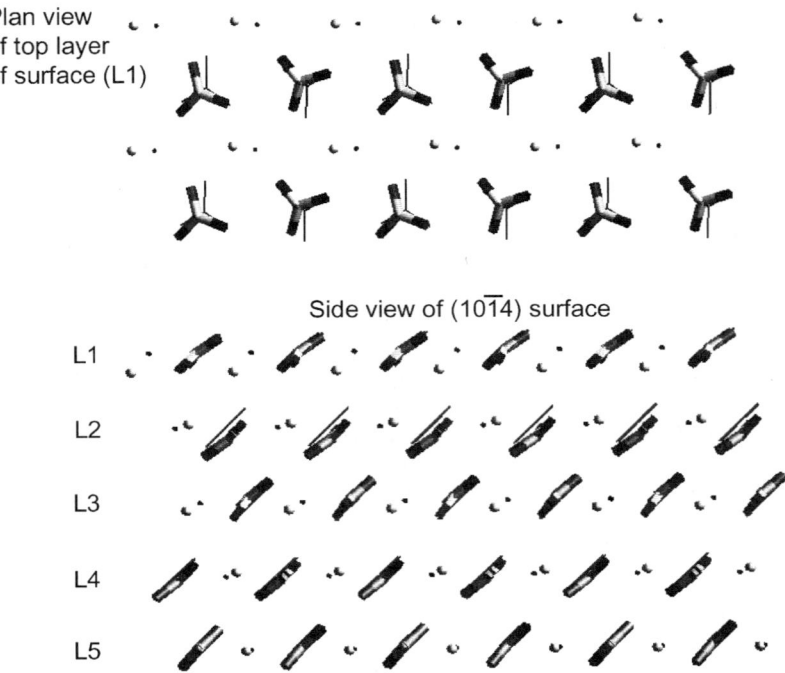

Figure 1. Plan and side views of the (10Ī4) surface of calcite as obtained using a shell model with full structural relaxation and energy minimization. The black lines for the carbonate groups and black dots for the calcium ions correspond to the bulk structure of calcite.

potential for the repulsive and van der Waals (attractive) short range interactions, and harmonic bond and bond angle terms to describe the covalent nature of the C-O bonds associated with the planar carbonate anion groups. Additionally, a shell model is used to describe the polarization of oxygen atoms [7] that will be important in controlling the relaxation response of the carbonate surface. We use the energy parameters obtained from a recent theoretical study of the bulk structures of calcite and several other metal carbonate phases [8]. Beyond the accurate simulation of bulk structures, the parameter set reproduces the bulk moduli (pressure derivatives), the calcite-aragonite transition, and the vibrational spectrum of calcite. We also incorporate a water model and water-carbonate interactions that have been previously used to simulate the (10Ī4) calcite surface [5].

Surface calculations were performed using the MARVINS code that calculates the surface energy for a given surface slice based on a multiple region approach [9]. The near surface environment of the carbonate (10Ī4) surface is represented by up to five layers that are allowed to be fully relaxed (free translation of all atomic coordinates) during the optimization of the total energy. A second region lies below this upper region and is constrained to the observed

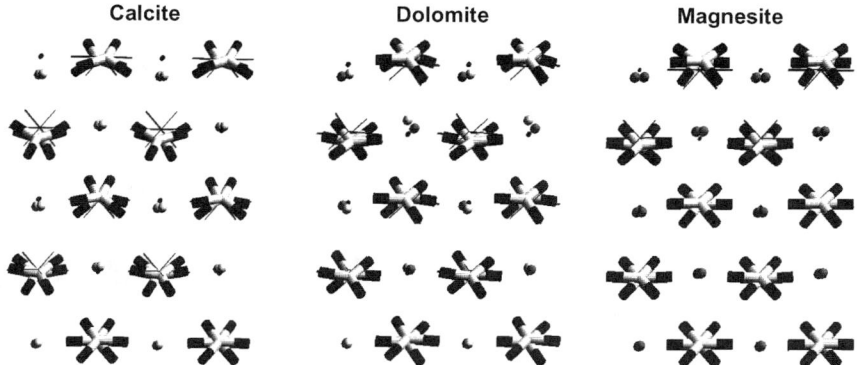

Calcite   Dolomite   Magnesite

Figure 2. *Comparison of relaxed (10$\bar{1}$4) surfaces for three carbonate compositions based on energy minimized structures with surface in contact with vacuum. The black lines for the carbonate groups and black dots for the metal ions correspond to the observed bulk structures.*

carbonate bulk structure. The MARVINS code incorporates a two-dimensional Ewald summation and optimization algorithm to obtain the minimum energy and surface structure. Simulations of wet surfaces are performed by the addition of a finite number of water molecules to the vacuum region above the upper carbonate surface region.

RESULTS AND DISCUSSION

The optimized structure for the (10$\bar{1}$4) surface of calcite as exposed to vacuum is presented in Figure 1. We determined that the choice of a five layer model for the upper surface region is sufficient for simulating the relaxation of the surface atoms. The positions of the fifth layer atoms are equivalent to those for the bulk calcite structure. The simulated calcite surface exhibits significant rotation and distortion of the carbonate groups accompanied by a displacement of the calcium ions downward from the surface. The second surface layer shows similar carbonate distortion but the calcium ions are significantly less displaced. The carbonate groups exhibit alternating layers of concave downward and concave upward distortions from the surface. Ultimately, the degree of distortion and displacement is reduced with depth. Surface structures for the (10$\bar{1}$4) surfaces of dolomite and magnesite are presented in Figure 2 along with that of calcite for comparison. The magnesite surface structure exhibits less carbonate group distortion and displacement of the magnesium ions than that observed for the calcite simulation. Relaxation of the magnesite surface appears to be relatively shallow; the observed bulk structure is obtained by the fourth layer. The mixed metal structure of dolomite exhibits a surface structure that has both carbonate group distortion and metal ion displacement, with the magnesium moving more in the topmost surface plane than calcium, but less in subsequent planes. The rotation of carbonate groups is significantly less than that observed for calcite.

The surfaces of calcite, dolomite, and magnesite were hydrated by placing a single water molecule on the surface and by covering with a monolayer of water. The molecules were

Figure 3. *Simulation result for relaxed (10$\bar{1}$4) surface of calcite with a monolayer of water.*

initially placed above the surface calcium or magnesium at an initial distance of 2.4 Å above the cation, and the carbonate surface and water were allowed to relax to their minimum energy configuration. Results for the simulation of the hydrous (10$\bar{1}$4) calcite surface are presented in Figure 3. The hydrated surface shows almost no difference from the bulk below the third layer while the degree of relaxation in the top two layers is substantially reduced compared to the dry surface. Water molecules nearest the calcite surface ultimately associate themselves with the calcium ions in order to complete the octahedral coordination of the metal ion with oxygens. Figure 4 provides a comparison of the variation of the torsional angle for the carbonate groups (O-C-O-O) as a function of layer position. Results are presented for both the dry and wet calcite surfaces. The greatest distortion occurs in the first layer for both simulations, but the trend with depth is vastly different with the wet surface exhibiting the bulk structural value by the fourth layer. A similar pattern emerges for the wet surfaces of dolomite and magnesite, where water stabilizes these surfaces by reducing the extent of surface relaxation and the surface energy. In the dry case, the surface energies are 0.322, 0.247 and 0.256 Jm^{-2} for calcite, dolomite, and magnesite respectively, and when fully hydrated with a monolayer of water, these energies are reduced to 0.315, 0.348 and 0.338 Jm^{-2}. The relative values for the surface energies suggest that the calcite surface is stabilized upon hydration while the corresponding surfaces for dolomite and magnesite are destabilized.

The results of our simulations of calcite are consistent with available experimental data [2,3] and with earlier simulations [4-6]. No comparable experimental data are available for dolomite or magnesite, although we are confident that the model used here is able to accurately reproduce the structure of these surfaces.

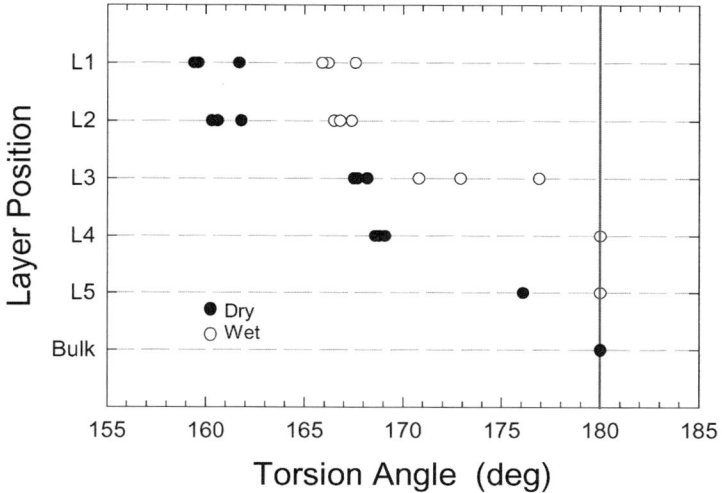

Figure 4. *Variation of the torsional angle for the carbonate groups as a function of layer depth for the simulated (10Ī 4) calcite surface in contact with vacuum (filled circles) and with a monolayer of water (open circles).*

CONCLUSIONS

Due to the limitations associated with experimental and spectroscopic observations, it is important to develop an atomistic model to help understand the energy and structural variations of calcite and other carbonate surfaces. We have developed a theoretical model that provides a description of the structure of the calcite, dolomite, and magnesite (10Ī4) surfaces. Surface relaxation occurs via carbonate group distortion and rotation, and combined with metal ion displacement, helps to reduce the surface energy. Simulations of the calcite surface in the presence of water indicate a significant reduction of distortion and surface energy.

Our future work involves collaboration with experimentalists at Argonne National Laboratory who are advancing the use of synchrotron sources for the analysis of mineral-water interfaces. In addition, we will extend our model to include a significant layer of water on the surfaces of these materials and assess the effects of temperature by running molecular dynamics simulations. In combination with accurate spectroscopic determinations of dry and wet carbonate surfaces, this theoretical approach will help to fully evaluate the structure and reactivity of an industrial and environmentally important class of materials.

ACKNOWLEDGMENTS

We would like to thank P. V. Brady for useful discussions regarding the applications of this study. This work was supported by the U.S. Department of Energy, Office of Basic Energy

Sciences, Geosciences Research Program, under contract DE-AC04-94AL85000 with Sandia National Laboratories. K. Wright is funded by the Royal Society under their University Research Fellowship program.

REFERENCES

1. P. van Cappellen, L. Charlet, W. Stumm and P. Wersin, *Geochim. Cosmochim. Acta*, **57**, 3505 (1993).
2. S. L. S. Stipp, *Geochim. Cosmochim. Acta*, **63**, 3121 (1999).
3. L. Cheng, N. C. Sturchio and M. J. Bedzyk, *Phys. Rev. B*, **61**, 4877 (2000).
4. J. O. Titiloye, N. H. de Leeuw and S. C. Parker, *Geochim. Cosmochim. Acta*, **62**, 2637 (1998).
5. N. H. de Leeuw and S. C. Parker, *J. Chem. Soc. Faraday Trans.*, **93**, 467 (1997).
6. N. H. de Leeuw, S. C. Parker and J. H. Harding, *Phys. Rev. B*, **60**, 13792 (1999).
7. B. G. Dick and A. W. Overhauser, *Physical Review*, **112**, 90 (1958).
8. D. K. Fisler, J. D. Gale and R. T. Cygan, *Am. Mineral.*, **85**, 217 (2000).
9. D. H. Gay and A. L. Rohl, *J. Chem. Soc. Faraday Trans.*, **91**, 925 (1995).

Mat. Res. Soc. Symp. Vol. 620 © 2000 Materials Research Society

The Kinetics of Calcite Growth: Interpreting Chemical Affinity-Based Rate Laws Through the Lens of Direct Observation

Henry. H. Teng[1], Patricia M. Dove[2] and James J. De Yoreo[3]
[1]Department of Geology
George Washington University
Washington, D.C. 20052
[2]Department of Geological Sciences
Virginia Polytechnic Institute and State University
Blacksburg, VA 24061
[3]Department of Chemistry and Materials Science
Lawrence Livermore National Laboratory
Livermore, CA 94550

ABSTRACT

Chemical affinity-based rate laws are used across the geochemical and materials communities to quantify mineral/material corrosion and growth kinetics. These rate expressions are founded in assumptions regarding reaction mechanism with little evidence for surface processes. Using Atomic Force Microscopy (AFM), this study demonstrates the dependence of growth kinetics upon the structures of dislocation sources. In situ observations show that the dominant mode of growth occurs by hillock development initiated at complex sources. Derivations of surface process-based rate expressions show a complex dependence of rate on chemical affinity. This dependence is approximated by second order affinity-based rate laws only under the special conditions that 1) growth proceeds by development of single sourced spirals and 2) growth occurs at very near equilibrium conditions where spiral formation is the only operative mechanism. This suggests that growth experiments that measure temporal changes in solution chemistry yield a composite rate that arises from the contributions of the different hillock types. Hence, chemical affinity-based rate laws do not generally give meaningful interpretations of growth mechanism. By combining direct observations with macroscopic methods that monitor temporal changes in solution chemistry, rate laws with greater predictive capabilities may be possible.

INTRODUCTION

The abundance of calcium carbonate minerals throughout natural and engineered earth systems has motivated investigations of calcite crystallization over the last century. It is widely recognized that an understanding of the kinetics and mechanisms governing growth is of first order importance for predicting mineralization and thus, acquiring the ability to control it. Advances in controlling or directing the growth of carbonate minerals hinge upon clarifying two uncertainties: (1) The dependence of growth mechanism upon supersaturation; (2) The microscopic surface processes that control the macroscopic manifestations of overall growth rates. Answers to these questions establish the knowledge base for constructing general models that quantify growth in complex mixtures of organic and inorganic constituents. They also establish relationships between microscopic processes and macroscopic growth rates determined by what are known as 'bulk' methods throughout the scientific literature. However, a review of the findings published to date suggests that these issues have not been adequately explored.

Growth studies conducted over a wide range of supersaturations have established our collective understanding of growth kinetics and mechanisms for calcite crystallization. Their findings can be summarized according to three experimental methods with different length-scales. At the largest scale, macroscopic studies have evaluated the rates and mechanisms of calcite crystallization using indirect methods that monitored changes in solution chemistry over the course of growth. These studies establish the dependence of growth kinetics upon chemical and physical parameters such as supersaturation, pH, P_{CO_2}, ionic strength, and temperature [1-5]. Growth mechanisms were postulated in these investigations based upon the rate of solution composition change with time. These 'bulk' kinetic measurements could not discern the possibly different contributions of unique crystal faces to the overall growth rate or relate solution chemistry to the range of microscopic growth processes occurring at mineral surfaces.

Three categories of kinetic models for crystallization have arisen from interpretations of solution composition data obtained by these indirect studies based upon (1) surface complexation [6,7]; (2) summation of the elementary reactions [2,3]; and (3) chemical affinity [8-12]. Surface complexation-based rate laws take into account the reactions involving surface speciation, while elementary reaction-based rate laws describe growth rate as a function of multiple elementary reactions. Affinity-based models, the most widely used of the three, are developed in terms of free energy changes, ΔG (chemical affinity) or $\exp(\Delta G/RT)$ (supersaturation index, Ω), of precipitation reactions [1,13-19] to yield two types of rate laws: linear and nonlinear with respect to ΔG. Linear rate laws have the following general form:

$$R_m = k_+[\exp(n \cdot G/RT)-1] \tag{1a}$$

where the net rate, R_m, (moles area^{-2} time^{-1}), is characterized by the rate constant of the forward reaction, k_+, (moles area^{-2} time^{-1}) and the free energy change of the overall reaction. R and T are the molar gas constant and temperature (K), respectively. The parameter n is a constant and has been assumed to contain information about the growth mechanism. For example, the kinetic behavior described by equation 1a with $n = 1$ has been attributed to adsorption-controlled growth (Nielsen, 1983). Expressions of this form with $n = 1/2$ and 1 were used to describe the precipitation kinetics of calcite [1,20].

Nonlinear rate laws are generally expressed as (Lasaga, 1981):

$$R_m = k_+[\exp(\cdot G/RT)-1]^n \tag{1b}$$

Theoretical models have been used to argue that second order equations of this form ($n = 2$) describe growth at screw dislocations by the spiral mechanism while higher order ones ($n = 2\sim3$) can be applied to growth at both screw and edge dislocations [22]. The kinetics of calcite precipitation has been fitted to second order expressions at ambient [15,23] and elevated temperatures [5]. Others have used an expression that combines equations 1a and 1b to describe calcite growth [24].

Together, these models delineate the currently accepted quantitative representations of calcite crystallization across a range of chemical compositions and conditions. However, these rate laws stand without substantiative confirmation of the actual growth processes that occur at mineral surfaces. This is a significant deficiency that constrains current models to provide only empirical and semi-quantitative representations of calcite precipitation kinetics. A closer analysis raises concerns about the relevance of these 'general' rate laws to calcite growth. The classic crystal growth theories predict that the growth rate of a spiral must be strongly controlled by the structure of dislocation sources [25]. Experimental observations have upheld these predictions [26-29]. In relating this fundamental aspect of crystal growth theory to accepted 'bulk' rate expressions, an important issue arises: Chemical affinity-based rate laws are not a function of dislocation source structures. The only measurable parameter in these expressions is Ω. Hence, rate laws represented by these expressions cannot reflect the controls of microscopic growth parameters (step velocity and slope of growth hillocks) on growth kinetics or reliably indicate the actual growth mechanism. Indeed, the broad range of overall growth rates that arise from the complexities of various dislocation sources naturally yield a quasi-parabolic behavior that has little mechanistic significance.

This study explores these concerns by comparing experimental evidence with the theoretical models to show where popular growth expressions are valid or break down. Using *in situ* Atomic Force Microscopy (AFM), we 1) determine the supersaturation range where growth occurs only by spiral formation; (2) determine the critical supersaturation value that marks the activation of growth by a surface nucleation mechanism; (3) measure step flow rates and the corresponding slopes of growth hillocks in well-characterized solutions; (4) calculate the overall or normal growth rates of single spirals using measurements of step speed and hillock slope; and (5) analyze the dependence of overall growth rate upon dislocation source structure. We show that spiral growth can be approximated by a second-order chemical affinity-based rate law only when supersaturation is *very* low and growth on a given crystal face is dominated by step flow from simple spirals. We also show that the kinetic data obtained in (4) represent only a minimum overall rate for pure spiral growth in calcite because growth generated by complex dislocation sources usually possesses a higher rate than that for single sources. A more complete discussion of these findings is found elsewhere [30].

EXPERIMENTAL

Calcite was grown on {104} faces of fragments cleaved from a crystal of optical-quality Iceland spar using supersaturated $CaCO_3$ solutions prepared from reagent grade sodium bicarbonate (NaHCO3, Aldrich®) and calcium chloride (CaCl2·H2O, Aldrich®) into deionized water. The ionic strength of each growth solution was fixed within a narrow range of 0.105-0.111 M using reagent grade NaCl (Baker®). The pH of all solutions was adjusted to 8.50 using 0.5 N NaOH prior to injection into the Fluid Cell of AFM.

The chemical speciation of each solution was assuming that the AFM fluid cell and input reservoir approximated a closed system. The activity ratio of Ca^{2+} to CO_3^{2-} was forced to equal 1.04±0.01 by adjusting the amount of NaHCO3 and CaCl2·H2O, and the Davies equation was used to correct for activity. The supersaturation, σ, was calculated by:

$$\sigma = \frac{\Delta\mu}{k_B T} = \ln(\frac{a}{a_e}) = \ln(\frac{a_{Ca^{2+}} a_{CO_3^{2-}}}{K_{sp}}) \qquad (2)$$

where $\Delta\mu$ is the chemical potential difference between $CaCO_3$ molecules in the aqueous and solid phases, k_B is the Boltzmann constant, a_i is the activity of ith species and K_{sp} is the solubility product of calcite at zero ionic strength. The value of $10^{-8.54}$ for K_{sp} at 25 °C was used to compute supersaturation [31]. We present results for $\sigma = 0.04$ to 1.4 ($\Omega = 1.04 - 4.06$).

In situ observations of calcite growth were made by Contact Mode® using a Nanoscope IIIa Scanning Probe Microscope (Digital Instruments) using a piezoelectric scanner with scan areas of up to 120x120 μm. Surfaces were imaged using commercially available Si3N4 cantilevers that have triangular tips with a length of 200 μm, a force constant of approximately 0.38 Newton m^{-1} and lever tip radii of approximately 30-50 nm.

The experiments were conducted by first imaging the mineral samples in air to locate a relatively flat area and to optimize image quality. The reactant solution was then input as a continuous flow through an AFM fluid cell with an internal volume of 50 μL using a syringe pump. Once in solution, the crystallographic orientation of an individual calcite fragment was established using methods described previously [32,33]. All images were collected using flow-through rates greater than 30 ml/hr. *In situ* measurements of temperature in this flow-through environment were 25° C. Data were collected by locating a single spiral hillock at each supersaturation. The spiral was allowed to grow for at least one hour in a supersaturated, near-equilibrium solution to ensure its quality and stability. The input solution was then changed to one with the desired σ and maintained for another 10 to 30 minutes (depending on imaging size and growth rate) while the step flow rate and geometry of the spiral adjusted to the new σ. Subsequently, a minimum of six images were captured in continuous mode for later measurements by image analysis. Measurements of step velocity, v_{s+} and v_{s-}, hillock slopes, p_+ and p_-, were made for the positive and negative directions.

RESULTS

Dependence of growth mechanism upon supersaturation

Layer growth was observed as the advancement of monomolecular (3Å) steps. These steps were generated by one or two mechanisms depending upon the surface structure and the saturation state of the input solution. At low values of σ ranging from 0 to 0.8, steps were initiated solely at dislocations, obvious crystal imperfections, and grain boundaries. Thus, crystal defect-originated growth by step flow from single and complex dislocation sources was the only growth mode. As shown in figure 1A-D and 2A, this mechanism resulted in several types of hillocks that became visible within a few minutes after the introduction of a supersaturated solution and new steps were formed only at defect sources that evolved into the apex of a hillock. Examination of more than 50 samples revealed that the most common types of hillocks (more than 80%) were multiple spirals generated by individual dislocations with multiple Burgers vectors (single sourced, multiple growth, figure 1A and B), dislocation groups (multi-sourced, multiple growth, figure 1D), and those that originated at obvious crystal imperfections (figure 1C). Single spirals (less than 20%) were the least common and often suppressed by multiple ones. In the areas where presumably no imperfections intersected the surface, or the high energy sources were overgrown by the newly crystallized layers, growth occurred only at the existing step edges and no new steps were generated.

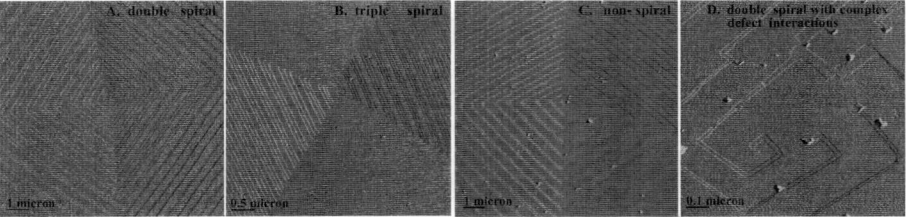

Figure 1. Growth hillocks on {104} faces of calcite [30]. (A) Growth generated by a dislocation with a Burgers vector of two results in a double spiral. (B) Growth generated by a dislocation with a Burgers vector of three results in a triple spiral. (C) Growth initiated at an obvious surface imperfection to yield a hillock with a growth unit of two mono-molecular layers. (D) A closeup view (1x1 micron) of growth initiated by a dislocation with a Burgers vector of two but complicated by large surface imperfections at result in a multi-sourced, multiple hillock.

When supersaturation exceeded approximately 0.8, steps were also generated by what appeared to be a homogeneous surface nucleation mechanism (figure 2B and 2C). However, AFM cannot distinguish nucleation on calcite lattice points (homogeneous) from nucleation at randomly distributed impurity sites or particles (heterogeneous). We refer to this type of mechanism to as homogeneous or two-dimensional surface nucleation.

Growth by surface nucleation occurred at randomly distributed two-dimensional nuclei on the substrate surface. Subsequent growth by this mechanism did not lead to hillock development. Nonetheless, growth originated from defects and by two-dimensional surface nucleation co-existed over the experimental supersaturation of $\sigma > 0.8$ as illustrated in figure 2B. When both mechanisms were operative, two-dimensional nuclei formed primarily within flat areas on the surface rather than on terraces of spiral hillocks. Increasing supersaturation resulted in faster nucleation rates and the development of more 2-D nuclei (figure 2C), indicating that growth by homogeneous surface nucleation was increasingly dominant with increasing supersaturation.

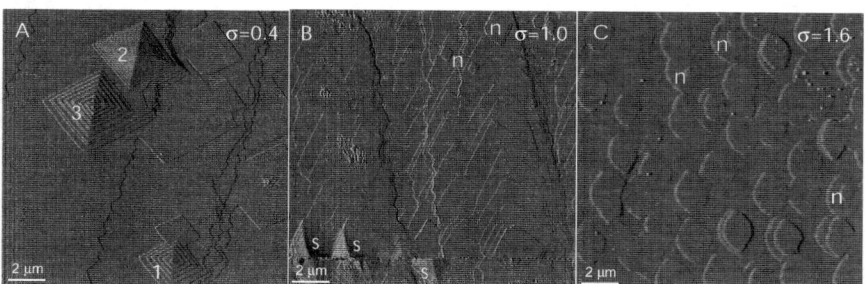

Figure 2. Fluid cell AFM images showing growth mechanisms on a (104) face at different supersaturation states. (A) Observations of spiral hillock formation collected at $\sigma = 0.4$ within minutes after input of growth solution. Three spirals are observed in the imaging area. Spirals 1 and 3 are single ones, and spiral 2 is a convolution of two double ones. In the area where dislocations are absent, growth occurred by the advancement of existing mono-molecular layers. (B) Co-existence of spiral growth (σ) and homogenous surface nucleation (n) at $\sigma = 1.0$. (C) The dominance of growth by two-dimensional surface nucleation was recorded at $\sigma = 1.6$ within tens of seconds after the input of solution. Two dimensional nuclei formed randomly on the surface. Continuous nucleation was also observed at several locations.

Growth rate of $\{10\,\overline{1}4\}$ faces for single spirals

The overall growth rate of a crystal face, or, the growth rate normal to the surface, R_m (length/time), is given by [25]:

$$R_m = pv_S \qquad (3)$$

Measurements of step velocities, v_S, and hillock slopes for single spirals, p, over the range of supersaturation used in this study were unique to the positive directions, $[\,\overline{4}41\,]_+$ and $[48\,\overline{1}\,]_+$, and the negative directions, $[\,\overline{4}41\,]_-$ and $[48\,\overline{1}\,]_-$. For each direction, v_S showed a complex dependence upon the deviation of equilibrium activity, $(a\text{-}a_e)$, where p scaled inversely with supersaturation, σ. Details of the dependence of direction-specific step velocity and slope on saturation state was discussed elsewhere [31,34]. The normal growth rate of a single spiral, R_m, was obtained using equation 3 and the measurements of v_S and p. Estimates of R_m increased from 10^{-10} - 10^{-8} mm s^{-1} over the experimental supersaturation range.

DISCUSSION

In situ AFM observations demonstrate that new steps are generated only at surface imperfections when supersaturation is low, but form by both crystal defects and two-dimensional surface nucleation as supersaturation increases. These observations are consistent with the predictions of classical BCF theories [25]. The observed evolution in mechanism with supersaturation has important implications for quantifying the overall growth rate, R_m, associated with microscopic surface processes and for interpreting the widely used macroscopic rate laws that are based upon chemical affinity.

Growth rates, R_m, controlled by simple and complex sources

The growth rate of crystal faces and, therefore, the overall growth rate of a crystal by a spiral mechanism is described by microscopic parameters v_s and p through equation 3 [25,26]. Because experimental observations show that growth of calcite at low supersaturations occurs by the formation of different types of hillocks, physically-representative rate expressions must account for the contributions of all of the important types of dislocation source structures to growth kinetics. This is difficult to achieve using chemical affinity based rate laws since the only variable in these expressions is the supersaturation ratio, Ω. This discussion examines the types of rate expressions that result from growth by three predominant and increasingly complex modes: single spirals, single-sourced multiple spirals (spirals generated by one dislocation with a Burgers vector > 2), and multi-sourced multiple spirals (spirals generated by multiple dislocations).

To begin, first recall that the dependence of step velocity upon solute activity in the absence of an impurity effect is given by:

$$v_s = \beta\omega(a-a_e)$$ (4)

where β is the kinetic coefficient (or rate constant) in length/time, ω is the specific molecular volume of crystal, and $(a-a_e)$ is the difference between actual and the equilibrium activities of the solute [35,36]. From equation 2, the solute activity in a solution can be expressed as:

$$a = a_e\exp(\frac{\Delta\mu}{k_BT}) = a_e\exp(\frac{\Delta G}{RT})$$ (5)

Substituting equation 5 into equation 4 gives the dependence of step velocity upon chemical affinity:

$$v_s = \beta\omega a_e[\exp(\frac{\Delta G}{RT})-1)]$$ (6)

Next, the terrace width, λ, of a single spiral can be estimated from the critical step length, L_c, by [25]:

$$\lambda = 4\Gamma L_c$$ (7)

where Γ relates to the delay of step advancement when step length is comparable to the magnitude of L_c and is approximately one for calcite [31]. In addition, L_c can be expressed as [25,26,31]:

$$L_c = \frac{2\omega\alpha}{k_BT\sigma}$$ (8)

where α is the step edge free energy (per unit length per unit height).

From this basis, the properties of the simplest type of spirals - a spiral generated by one dislocation with a Burgers vector of one (figure 1A) - can be quantified. The hillock slope, p, of such spirals can be obtained by combining the relation $p = h/\lambda$ with equations 7, and 9 into the expression:

$$p = \frac{hk_BT}{8\alpha\omega}\left(\frac{\Delta G}{RT}\right) \tag{9}$$

which indicates that hillock slopes have a positive correlation with solution supersaturation. Notice that the definition

$$\sigma \equiv \Delta G/RT \tag{10}$$

was used to develop equation 9. It follows from equation 3 that, in the absence of impurities, the overall growth rate for the simplest type of spiral growth is the product of equation 6 and 9:

$$R_m = \frac{a_e\beta hk_BT}{8\alpha}[\exp(\tfrac{\Delta G}{RT})-1](\tfrac{\Delta G}{RT}). \tag{11}$$

This is illustrated in figure 3A that shows R_m has a superlinear dependence upon supersaturation.

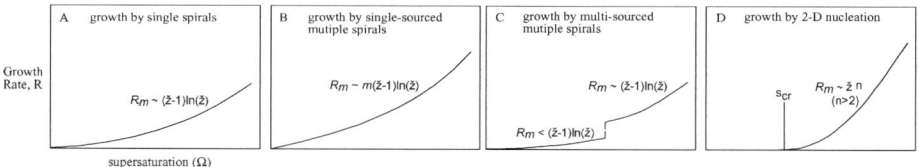

Figure 3. Schematic representation that shows the dependence of overall growth rate on supersaturation index (Ω) in different growth modes. All plots are referenced to the same scale. (A) Growth controlled by single spirals. (B) Growth controlled by single-sourced multiple spirals. (C) Growth controlled by multi-sourced multiple spirals. (D) Growth controlled by two-dimensional surface nucleation. When spiral growth is the only mechanism, the minimum growth rate may be considered as that given for single spirals. When growth is controlled by both spiral and surface nucleation mechanisms, the overall rate is the sum of the contributions from each growth mode.

A more complex, second type of spiral growth can occur when the dislocation source has a Burgers vector of m (figure 2A-B). In this case, the growth unit is multiple steps of m. Hence, equation 9 becomes

$$p = \frac{mhk_BT}{8\alpha\omega}\left(\frac{\Delta G}{RT}\right) \tag{12}$$

and R_m can be estimated by:

$$R_m = \frac{ma_e\beta hk_BT}{8\alpha}[\exp(\tfrac{\Delta G}{RT})-1](\tfrac{\Delta G}{RT}). \tag{13}$$

This is shown in figure 3B where the dependence of R_m on supersaturation is m times greater than for a single spiral.

A third, more complicated type of spiral growth occurs when the source is a group of dislocations. It has been demonstrated that, for a group of screw dislocations, when the separation between them is smaller than one half of the step width generated by individual dislocations, they co-operate and form a multi-sourced multiple spiral [25]. Consider a group of m screw dislocations of the same rotation direction lying on a line of length Λ. Assuming that the separations between these dislocations are less than $\lambda/2$ generated by individual dislocations, a multi-sourced multiple spiral hillock develops (figure 2D). The magnitude of p in this case is controlled by m, Λ, and L_c through [26-28]:

$$p = \frac{mh}{4\Gamma L_c + 2\Lambda} .$$

(14)

Combining equations 8 and 10 and then substituting into equation 14 yields the dependence of hillock slope upon chemical affinity for a multi-sourced spiral:

$$p = \frac{mhk_B T(\frac{\Delta G}{RT})}{8\omega\alpha + 2\Lambda k_B T(\frac{\Delta G}{RT})} .$$

(15)

Assuming step velocity is independent of p, the product of equations 6 and 15 gives an expression for the rate of a multi-sourced multiple spiral growth:

$$R_m = \frac{a_e \omega h \beta mk_B T(\frac{\Delta G}{RT})}{8\omega\alpha + 2\Lambda k_B T(\frac{\Delta G}{RT})}[\exp(\frac{\Delta G}{RT})-1] .$$

(16)

equation 15 can also be applied to situations where the dislocations are not arranged in a straight line. In this case, 2Λ represents the perimeter of the area occupied by these dislocations [26].

It is important to also note that, for this type of spiral growth, rate can have an apparent discontinuous dependence upon supersaturation. As pointed out earlier in this section, a multi-sourced spiral forms from a group of dislocations only when $\lambda/2 > \Lambda$ [25]. Because terrace width, λ, scales inversely with supersaturation (equations 7 and 8), this complex spiral is stable at low supersaturations where step width is large. With increasing supersaturation and decreasing step width, the spiral decomposes into a number of single spirals that are generated by each individual dislocation when $\lambda/2$ becomes smaller than Λ. When this occurs, the hillock slope will, in general, increase discretely from the value given by equation 15 to that by equation 9. Hence, the overall growth rate, R_m, is no longer a monotonic function of supersaturation, as illustrated in figure 6C and demonstrated by experiment [28].

Relationships between process-based and affinity-based rate expressions

Because direct observations of calcite growth show that step flow originates from predominantly complex sources and, to a lesser extent, from simple sources, the net rate of spiral growth across one surface - natural or synthetic, is necessarily a complex composite of contributions from several R_m expressions (e.g. equations 11, 13, and 16), each of which has a unique dependence upon saturation state. Hence, the question arises: Are the reported macroscopic, chemical affinity-based rate laws related to microscopic processes occurring at the growing surface? The obvious conclusion from the preceding analysis is that the growth rates determined by measuring temporal changes in bulk solution compositions must also be composites. If no, are there conditions where comparisons of the two types of rate laws are meaningful?

To answer these questions, first consider equation 16, the more complicated rate expression for spiral growth. This equation reduces to the simpler equation 13 when m

dislocations overlap, i.e. $\Lambda = 0$, to form a single source with a Burgers vector of m. equation 13 further reduces to equation 11, the rate expression for the simplest type of spiral growth, when the Burgers vector of the dislocation becomes one ($m = 1$). Equation 11 can be further simplified to:

$$R_m \approx \frac{a_e \beta h k_B T}{8\alpha}[\exp(\frac{\Delta G}{RT})-1]^2 \tag{17}$$

when supersaturation is very low or ($\Delta G/RT$) << 1. The form of this expression is analogous to a second order chemical affinity-based rate law (equation 1b with $n = 2$) provided that $\frac{a_e \beta h k_B T}{8\alpha}$ represents the rate constant, k_+.

The above comparison suggests that spiral growth is approximated by a second order chemical affinity based-rate law only when (1) the supersaturation is so low that spiral formation is the only growth mechanism and (2) the formation of simple spirals dominates the growth of all crystal faces. Direct observations (figure 2) reported in this study show that even in the supersaturation range where spiral formation is the only growth mode, the second requirement fails because multiple spirals are far more commonly observed than single ones. This suggests that even when growth occurs at low supersaturation and the spiral mechanism is the only valid growth mode, still no single kinetic expression can describe growth because the dislocation source structures may be sample specific. Rather, R_m measured at each supersaturation for pure spiral growth must be within a range that is defined by equations 11 and 16 for the lower and upper boundaries (figure 5A-B, respectively). The lower boundary for the growth of calcite {104} faces in this study is constrained by experimental measurements [30]. Finally, in the presence of obviously large surface defects such as pits, fractures, and grain boundaries (figure 2C-D), the determination of R_m can be further complicated because the growth is controlled by the specific characteristics of each source. Most significantly, if the source is at a gross defect where steps must circumscribe a structural imperfection of a perimeter Π, the term 2Λ in equations 14 to 16 is replaced by Π. The result is that as σ increases and L_c decreases, the second term in the denominator becomes dominant, the slope becomes independent of supersaturation and R_m becomes a linear function of Ω.

When supersaturation becomes high enough that two-dimensional surface nucleation mechanism becomes operative (figure 3D), the contributions of two-dimensional nuclei formation to total R_m must be added. The macroscopic growth rate by a two-dimensional nucleation mechanism has an exponential dependence upon supersaturation through the expression

$$R_m = 1.137h(Iv_s^2)^{1/3} \tag{20}$$

where h is the step height, I the nucleation frequency (area^{-2} time^{-1}) [37]. Because I increases exponentially with increasing supersaturation, contributions of the two-dimensional surface nucleation mechanism are increasingly important with increasing supersaturation and quickly become the dominant component in the overall growth rate. In the supersaturation range where spiral and two-dimensional nucleation mechanisms co-exist ($\sigma > 0.8$ or • > 2.2 for this study), the overall growth rate will be the sum of all growth modes illustrated in figure 3A-D.

CONCLUSIONS

The clear conclusion from experimental observations and this analysis is that attempts to relate affinity-based rate laws to microscopic growth mechanism are of limited value. Macroscopic rate laws derived from bulk experiments represent the composite average of quite different growth rate expressions for the unique reactivities of many different crystal facets and surface structures. However, affinity-based rate laws comprise most of our quantitative knowledge regarding calcite growth kinetics and will continue to be employed by the geological and engineering communities.

Results of this study demonstrate the need to apply caution when deducing growth mechanisms and rate laws from temporal changes in bulk solution chemistry. Further, interpretations of growth mechanism using data collected from these indirect methods are particularly hazardous without direct evidence for the growth processes that are occurring at the mineral surfaces. The analysis in this study suggests that popular 'rate laws' are empirical, at best. By combining direct observations with these methods, improved rate laws with greater predictive capabilities may be possible.

ACKNOWLEDGEMENTS

This work was supported by the Chemical Sciences, Geosciences and Biosciences Division, Office of Basic Energy Sciences, U.S. Department of Energy through grant number DE-FG05-95-ER14517 and was performed under the auspices of Lawrence Livermore National Laboratory under contract W-7405-Eng-48. We thank P. Vekilov for helpful comments.

REFERENCES

1. Nancollas G. H. and Reddy M. M. (1971) The crystallization of calcium carbonate II: Calcite growth mechanism. *J. Colloid Interface Sci.* **37**, 843-830.
2. Plummer L. N., Wigley T. M. L., and Parkhurst D. L. (1978) The kinetics of calcite dissolution in CO_2-water systems at 5-60°C and 0.0-1.0 atm CO_2. *Amer. J. Sci.* **278**, 179-216.
3. Busenberg E. and Plummer L. N. (1986) A comparison study of the dissolution and crystal growth kinetics of calcite and aragonite. *USGS Bull.* **1578**, 139-168.
4. Christoffersen J. and Christoffersen M. R. (1990) Kinetics of spiral growth of calcite crystals and determination of the absolute rate constant. *J. Crystal. Growth* **100**, 203-211.
5. Shiraki R. and Brantley S. L. (1995) Kinetics of near-equilibrium calcite precipitation at 100°C: An evaluation of elementary reaction-based and affinity based rate laws. *Geochim. Cosmochim. Acta* **59**, 1457-1471.
6. Nilsson Ö and Sternbeck J. (1998) A mechanistic model for calcite crystal growth using surface speciation. *Geochim. Cosmochim. Acta* **63**, 217-226.
7. Arakaki, T. and Mucci, A. (1995) A continuous and mechanistic representation of calcite reaction-controlled kinetics in dilute solutions at 25°C and 1 Atm total pressure. *Aquatic Geochemistry*, **1**, 105-130.
8. Smallwood P. V. (1977) Some aspects of the surface chemistry of calcite and aragonite. *Colloid and Polyner Sci.* **255**, 994-1000.
9. Reddy M. M. (1977) Crystallization of calcium carbonate in the presence of trace concentrations of phosphorous-containing anions. *J. Crystal Growth* **41**, 287-295.
10. Reddy M. M. and Gaillard W. D. (1980) Kinetics of calcium carbonate (calcite)-seeded crystallization: Influence of solid/solution ratio on the reaction rate constant. *J. Colloid and Interface Science*, **80**, 171-178.
11. Reddy M. M. (1988) Physical-Chemical mechanisms that affect regulation of crystallization. In *Chemical Aspects of Regulation of Mineralization* (eds. C. S. Sikes and A. P. Wheeler), 3-8, University of South Alabama Pub. Ser., Mobile, Alabama.
12. Compton R. G. and Daly P. J. (1987) The dissolution/precipitation kinetics of calcium carbonate: An assessment of various kinetic equations using a rotating disk method. *J. Colloid. Interface Sci.* **115**, 493-498.
13. Reddy M. M. and Nancollas,G. H. (1971) The crystallization of calcium carbonate I.

Isotopic exchange and kinetics. *J. Colloid and Interface Science*, **36**, 166-172.

14. Morse J. W. (1978) Dissolution kinetics on calcium carbonate in sea water VI: The near equilibrium dissolution kinetics of calcium carbonate-rich deep sea sediments. *Amer. J. Sci.* **278**, 344-353.

15. House W. (1981) Kinetics of crystallisation of calcite from calcium bicarbonate solutions. *J. Chem. Soc. Faraday Trans.* **77**, 341-359.

16. Nielsen A. E. (1983) Precipitates: formation, coprecipitation, and aging. In *Treatise on Analytical Chemistry (eds.* I. M. Kolthoff and P. J. Elving), 269-374, Wiley, New York.

17. Mucci A. (1983) The solubility of calcite and aragonite in seawater at various salinities, temperatures, and one atmosphere total pressure. *Amer. J. Sci.* **283**, 780-799.

18. Mucci A. and Morse, J. W. (1983) The incorporation of Mg and Sr into calcite overgrowths: Influences of growth rate and solution composition. *Geochim. Cosmochim. Acta* **47**, 217-233.

19. Mucci A. (1986) Growth kinetics and composition of magnesian calcite overgrowths precipitated from seawater: Quantitative influence of orthophosphate ions. *Geochim. Cosmochim. Acta* **50**, 2255-2265.

20. Kazmierczak, T. F., Tomson, M. B., and Nancollas, G. H (1982) Crystal growth of calcium carbonate: A controlled composition kinetic study. *J. Phys. Chem.* **86**, 103-107.

21. Lasaga A. C. (1981) Transition state theory. In *Kinetics of Geochemiscal Processes* (eds. A. C. Lasaga and R. J. Kirkpatrick); *Rev. Mineral.*, 135-169.

22. Blum A. E. and Lasaga A. C. (1987) Monte Carlo simulations of surface reaction rate laws. In *Aquatic Surface Chemistry* (ed. W. Stumm), 255-292, Wiley, New York.

23. Reddy, M. M. and Nancollas, G. H. (1973) Calcite crystal growth inhibition by phosphates. *Desalination* **12**, 61-73.

24. Inskeep W. P. and Bloom P. R. (1985) An evaluation of rate equations for calcite precipitation kinetics at Pco_2 less than 0.01 atm and pH greater than 8. *Geochimica et Cosmochimica Acta* **49**, 2165-2180.

25. Burton W. K., Cabrera N., and Frank F. C. (1951) The growth of crystals and the equilibrium structure of their surfaces. *Royal Soc. London Philos. Trans.* **A243,** 299-358.

26. Rashkovich, L. N. (1991) *KDP-family single crystals.* 100-165, IOP Pub., Norfold, England.

27. Vekilov P. G. and Kuznetsov Yu. G. (1992) Growth kinetics irregularities due to changed dislocation source activity: (101) ADP face. *J. Crystal Growth*, **119**, 248-260.

28. Vekilov P. G. and Rosenberger F. (1996) Dependence of lysozyme growth kinetics on step sources and impurities. *J. Crystal Growth*, **158**, 540-551.

29. Land T. A., De Yoreo, J. J., and Lee, J. D. (1997) An *in situ* AFM investigation of canavalin crystallization kinetics. *Surf. Sci.* **384**, 136-155.

30. Teng H. H., Dove P. M., Orme C. A., and DeYoreo J. J. (2000) Kinetics of calcite growth: Surface processes and relationships to macroscopic rate laws. *Geochimica et Cosmochimica Acta*, **64**, in press.

31. Teng H. H., Dove P. M., Orme C. A., and DeYoreo J. J. (1998) Thermodynamics of calcite growth: Baseline for understanding biomineral formation. *Science*, **282**, 724-727.

32. Teng H. H. and Dove P. M. (1997) Surface site-specific interactions of aspartate with calcite during dissolution: Implications for biomineralization. *Amer. Mineral.* **82**, 878-887.

33. Stipp S. L., Eggleston C. M., and Nielsen, B. S. (1994) Calcite surface structure observed at microtopographic and molecular scales with atomic force microscopy (AFM). *Geochimica et Cosmochimica Acta*, **58**, 3023-3033.

34. Teng H. H., Dove P. M., and DeYoreo J. J. (1999) Reversed calcium carbonate morphologies induced by microscopic growth kinetics: Insight into biomineralization. *Geochimica et Cosmochimica Acta*, **63**, 2507-2512.

35. Chernov A. A. (1961) The spiral growth of crystals. *Soviet Phys.* **4**, 116-148.

36. Chernov A. A. and Komatsu H. (1995) Topics in crystal growth kinetics. In Science and Technology of Crystal Growth (eds. J. P. van Eerden and O. S. L. Bruinsma), 67-80, Kluwer Acad. Pub., Amsterdam.

37. Van der Eerden J.P. (1993) Crystal Growth Mechanisms. In Handbook of Crystal Growth (ed. D.J.T. Hurle) **1A**, 307-475.

37. Van der Eerden J.P. (1993) Crystal Growth Mechanisms. In Handbook of Crystal Growth (ed. D.J.T. Hurle) **1A**, 307-475.

Mat. Res. Soc. Symp. Vol. 620 © 2000 Materials Research Society

Glass-Crystal Boundaries In Liquid-Phase Sintered Ceramics

N. Ravishankar and C. Barry Carter
Department of Chemical Engineering and Materials Science, University of Minnesota
421 Washington Ave SE, Minneapolis, MN 55455-01432

ABSTRACT

The interface between dewet glass droplets and the free surface of a crystal and the interface between the intergranular glass and adjoining crystalline grains have been examined with particular emphasis on the influence of the crystallography of the free surface and the grain boundary. The wetting of liquid on the free surface has been shown to depend on the surface structure. The migration of boundaries containing a liquid phase has been studied. The migration is initiated by the difference in surface energy of the bounding planes. Faceting of the grain boundary planes has been examined. It is proposed that the boundary migrates by the motion of the facets.

INTRODUCTION

Liquid-phase sintering (LPS) is a common route for processing many of the commercial ceramic materials [1]. The liquid phase that forms at the annealing temperature enhances the kinetics of the process, thus making it commercially viable. The liquid that aids in the sintering process is usually retained as a glassy phase in the microstructure. The glass is distributed along various grain boundaries and multi-grain junctions. It has been reported for oxide ceramics that there is a variation in the thickness of the intergranular glass layer depending on the energy of the grain boundary [2, 3] while it has been reported to be independent of boundary orientation in case of Si_3N_4 [4]. In either case, the glass-crystal interface thus formed controls the properties of the sintered product and thus it is important to understand the structure and behavior of this interface. In spite of its importance, very little is known about this interface.

The present paper reviews the glass-crystal interface with particular reference to the liquid-phase sintering process. In particular, the interface between dewet glass droplets and free surface of the crystal and the migration of grain boundaries containing intergranular glass layer will be discussed.

The interface between the glass and the free surface of a crystalline material provides a basis for understanding the interface between intergranular glass and the crystalline grains. Detailed examination of the solid/liquid/vapor (SLV) triple junction reveals many important aspects of the liquid-solid interactions [5-8]. The wetting behavior of liquids on reconstructed high-index surfaces provides useful insight on the importance of surface energy on wetting behavior. The faceting of higher energy surfaces in alumina into lower energy facets on annealing at high temperatures is well-characterized [9, 10]. It has been demonstrated that the faceting behavior of the free surface in contact with a glassy phase is significantly enhanced compared to the faceting of the clean free surface [11, 12]. This difference has important implications with respect to the faceting of grain boundaries which contain an intergranular glassy phase. An understanding of the interface between intergranular glass and the crystalline grains is directly relevant in controlling the properties of liquid-phase sintered materials. In many cases, it is desirable to eliminate the grain boundary glass. Crystallization of the glass or the exudation of glass to free surface are two possibilities. The importance of crystallography on these processes has been demonstrated [13-15].

In the present paper, solid-liquid interaction in the alumina/anorthite system has been studied. In order to bring out the influence of crystallography, samples with well-defined starting states are used. This effect is pronounced in the case of an anisotropic crystal like alumina. The use of single crystal surfaces of known orientations or bicrystals with controlled misorientation has been shown to be very effective in this regard [12, 13].

EXPERIMENT

Pulsed-laser Deposition (PLD) is used to deposit thin films (~100 nm) of anorthite ($CaAl_2Si_2O_8$) glass on single–crystal sapphire substrates. The bicrystals are engineered by bonding the glass-coated single crystal of sapphire to a blank piece of sapphire under an applied pressure. The crystallography is tailored by orienting the sapphire crystals before hot-pressing. The bicrystals are cut using a diamond saw and polished using the tripod polishing technique. This process gives a grain boundary that is perpendicular to the polished free surface. The bicrystal assembly is annealed at 1650°C for 2 h. A field-emission SEM operated at 5 kV in the secondary-electron (SE) mode has been used for microstructural investigation. Samples for the SEM analysis were coated with a thin layer (~1-2 nm) Pt to avoid charging.

RESULTS AND DISCUSSION

Fig. 1 is an SE image of a dewet droplet on a high-index surface of alumina. The high-index surface is unstable at the annealing temperature and reconstructs into a hill-and-valley structure with two different types of facets. The morphology of the droplet deviates significantly from the circular cross-section that is expected for a liquid droplet on a flat substrate.

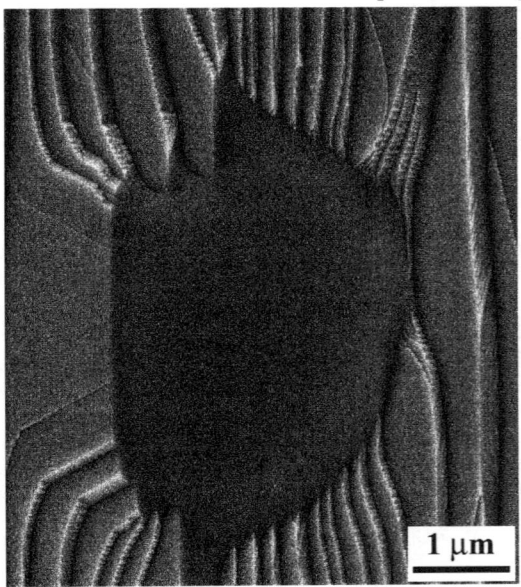

Fig. 1 SE image showing a dewet droplet of anorthite on a reconstructed surface of alumina. The sample was annealed at 1650°C for 2h. The unstable surface of alumina reconstructs into a hill-and-valley structure on annealing. The morphology of dewet droplet on the stepped surface depends on the surface energies and distribution of the individual facets.

The solid/liquid/vapor triple junction is modified at the regions where the facets are present. It is seen that the liquid spreads along one type of facet on the surface. The wetting behavior of anorthite liquid on various single-crystal surfaces of sapphire has been studied in great detail . It has been shown that the morphology of dewet droplets depends on the surface energy and the surface structure [5-8]. In particular, the faceting of an unstable surface affects the morphology of dewet droplets significantly.

Fig.2 is an SE image from a reconstructed surface that clearly brings out the effect of the surface energy of the facets. The liquid in this case spreads significantly along the narrow facets on the surface due to the fact that the solid/liquid interfacial energy is lower for this facet. The non-circular cross-section is formed in order to maximize contact with the favorable facet. Thus, the droplets develop a D-shape with flat section on one side (evident in both Fig. 1 and Fig. 2) in order to maximize contact with the favorable facet while on the other side, the cross-section is tangential to the unfavorable facet in an effort to minimize the contact. From the anorthite/alumina phase diagram, there is a significant amount of alumina that dissolves in the liquid at the annealing temperature. The precipitation of alumina takes place on cooling the sample. The precipitation also proceeds in such a way as to minimize the solid/liquid interfacial energy.

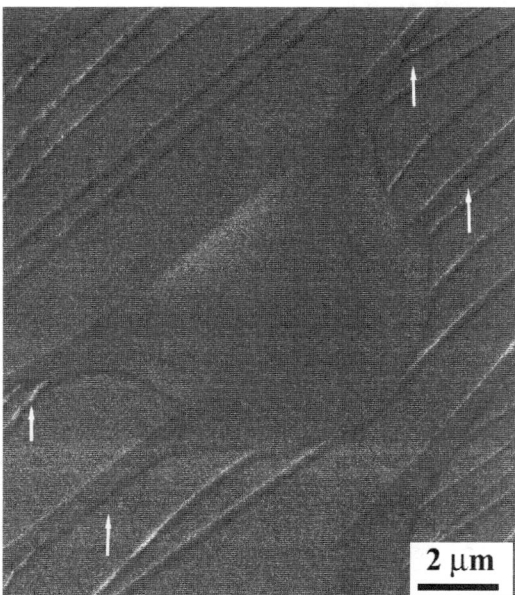

Fig. 2. SE image of a dewet droplet on a reconstructed surface of alumina. The liquid that forms at the annealing temperature preferentially wets one of the facets in the faceted structure (indicated by arrows). This observation brings out the importance of crystallography on the wetting behavior.

Fig. 3 is an SE image from a bicrystal bounded by the m-plane and c-plane of alumina (m|c bicrystal) that has been annealed at 1650°C for 2h. The lower portion of the boundary has migrated to the right from its initially straight position. The initial state and the configuration after annealing have been shown in the schematic to the right. The boundary groove that forms takes a longer time to heal compared to the time for its motion and hence the initial position of the boundary is evident from the surface groove that it leaves behind [16]. The free surface

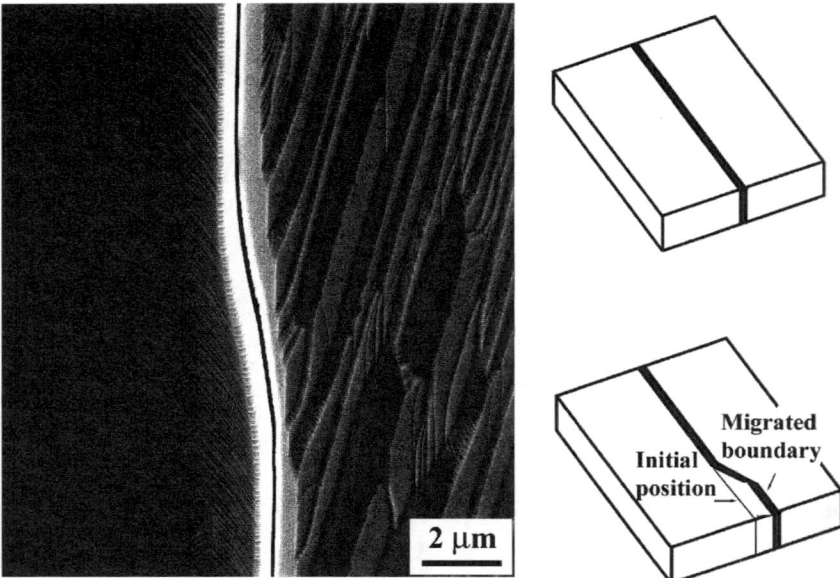

Fig. 3 SE image from an m|c bicrystal showing the migration of the grain boundary. The initial configuration and the migrated bounary have been indicated in the schematic to the right. The initial position of the boundary is evident from the interaction of the boundary groove with the free surface

to the right contains dewet droplets while the surface to the right does not. The exudation of liquid silicate from grain boundaries has been discussed in terms of the driving force for the process [13, 15]. The migration process can be understood in terms of the solution-reprecipitation process taking place at high temperatures. Most of the liquid-phase sintering systems have an eutectic phase diagram with a small solubility in the solid phase. Sintering is usually carried out at temperatures slightly above the eutectic temperature. At this temperature, there is a considerable amount of solid that dissolves in the liquid and also a reprecipitation of the solid which takes place simultaneously. This solution-reprecipitation at the solid-liquid boundaries leads to densification. The driving force for solution-reprecipitation is provided by a difference in curvature between adjacent grains. Larger grains grow at the expense of smaller grains adjacent to them by an Ostwald ripening process. The alumina/anorthite pseudo-binary phase diagram is an eutectic diagram with no terminal solid solubility. Alumina dissolves in anorthite liquid at the temperature of annealing. This process continues until the liquid composition attains the composition given by the phase diagram. The equilibrium is a dynamic one and there is simultaneous dissolution and precipitation of alumina that takes place. There is no curvature effect in the present case as the bounding planes are polished flat planes. However, there is a difference in the surface energies of the two planes and thus the solution and precipitation rates from the two surfaces are expected to be different. This may lead to a concentration gradient in the liquid that drives the boundary in a particular direction. The concentration gradient is essentially developed by a fluctuation process that acts as the nucleation step. Once nucleation has occurred, the growth is rapid and self-sustained.

Although the preceding section gives a phenomenological picture of the migration process, the exact mechanism is not discussed. Fig. 4 is an SE image from the m|c bicrystal showing the details of faceting at the interface in the migrated region of the boundary. The steps on the lower portion of the boundary (c-plane) can be clearly seen. The density of steps changes in order to accommodate the inclination of the boundary from the initial straight position. The steps are presumably responsible for the migration of the boundary.

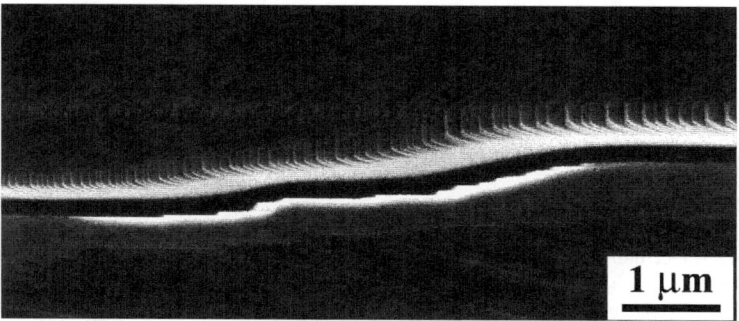

Fig. 4 SE image from a m|c bicrystal annealed at 1650°C for 2h. The faceting of the boundary plane is evident. The boundary migrates by the motion of the facets.

Fig. 5 is a diffuse dark-field TEM image from an m|a bicrystal showing the details of faceting at the interface [17]. The m-plane (lower bounding plane) facets in to two sets of planes as indicated. The glass layer (that appears bright) is almost absent in some regions of the interface (i.e) there are regions of crystal/crystal contact at this interface. The use of TEM provides additional valuable information on the microscopic aspects of faceting.

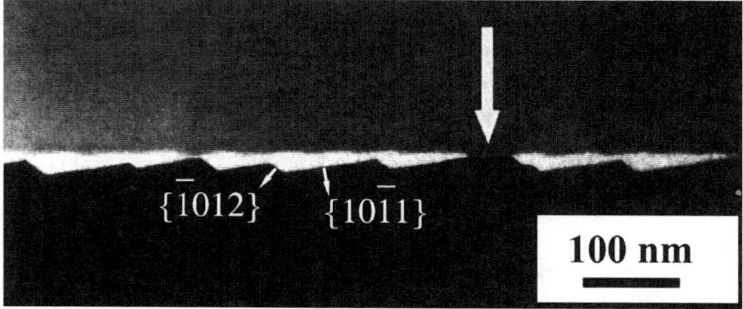

Fig. 5. Diffuse dark-field TEM image from a m|a bicrystal showing the faceting of the interface [17]. The glass appears bright in this image. Regions of flattened contact form at this interface.

CONCLUSIONS

The use of single crystals and bicrystals of known misorientations clearly reveals the influence of crystallography on the solid-glass interactions on a microscopic scale. Wetting of liquids on faceted surfaces depends strongly on the surface energies of the facets formed. The liquid spreads along favorable facets and the morphology of the liquid is distorted to maximize the contact with the favorable facets. Boundary migration in bicrystals proceeds by solution-reprecipitation events even in the absence of curvature driving force. Detailed study of the faceting leads to an understanding of the mechanism of migration.

ACKNOWLEDGMENTS

This research is supported by the U. S. Department of Energy under Grant No. DE-FG02-92ER45465. The authors would like to acknowledge Prof. Stan Erlandsen for access to the FESEM and Chris Frethem for technical assistance.

REFERENCES

1. R.M. German, *Sintering Theory and Practice,* John Wiley and Sons, Inc., New York, 1996.
2. D.R. Clarke, Ann. Rev. Mater. Sci. **17**, pp. 57-74 (1987).
3. D.R. Clarke in *Intergranular Phases in Polycrystalline Ceramics,* edited by L.-C. Dufour, C. Monty, and G. Petot-Ervas (Kluwer Academic Publishers, Dordrecht, 1989), pp. 57-79.
4. M.K. Cinnibulk, H.-J. Kleebe, G.A. Schneider, and M. Ruhle, J. Am. Ceram. Soc **76**, pp. 2801-808 (1993).
5. N. Ravishankar, S. Ramamurthy, H. Schmalzried, and C.B. Carter in *Wetting of Anorthite Liquid on m-Sapphire Substrates,* edited by (Microscopy and Microanalysis '99Springer, Portland, OR, pp. 812-813.
6. S. Ramamurthy, B.C. Hebert, and C.B. Carter, Phil. Mag. Lett. **72**, pp. 269-75 (1995).
7. S. Ramamurthy, B.C. Hebert, C.B. Carter, and H. Schmalzried, Mat. Res. Soc. Symp. Proc. **398**, pp. 295-300 (1996).
8. S. Ramamurthy, C.B. Carter, and H. Schmalzried, Phil. Mag (in press), (2000).
9. J.R. Heffelfinger, M.W. Bench, and C.B. Carter, Surf. Sci. Lett. **343**, pp. 1161-66 (1995).
10. J.R. Heffelfinger and C.B. Carter, Surf. Sci. **39**, pp. 188–200 (1997).
11. Y.K. Simpson and C.B. Carter, J. Am. Ceram. Soc. **73**, pp. 2391-2398 (1990).
12. D.W. Susnitzky and C.B. Carter, J. Am. Ceram. Soc. **73**, pp. 2485-93 (1990).
13. N. Ravishankar and C.B. Carter in *Migration of Silicate Liquids out of Grain Boundaries in Ceramics,* edited by (Microscopy and Microanalysis '99) Springer, Portland, OR, pp. 800-801.
14. M.P. Mallamaci, J. Bentley, and C.B. Carter, Acta Mater. **46**, pp. 283-303 (1998).
15. N. Ravishankar and C.B. Carter, submitted to J. Am. Ceram. Soc, (2000).
16. W.W. Mullins, Acta Metall. **6**, pp. 414-427 (1958).
17. M.P. Mallamaci and C.B. Carter, Acta Materialia **46**, pp. 2895-2907 (1997).

Growth of Organic Films and
Supramolecular Solids

Mat. Res. Soc. Symp. Vol. 620 © 2000 Materials Research Society

Ordered Thin Films of Perylenetetracarboxylicdianhydride-bisimide and bis-(N-alkyl)-Quinacridone Dyes

Andrew Back,[1,2] Dana Alloway[1], Derck Schlettwein[1,3], Brook Schilling[1,4],
J.-F. Wang[1], Mike Carducci[1], Neal R. Armstrong[1]
[1]Department of Chemistry, University of Arizona, Tucson, Arizona 85721
[2] Physical Electronics Corp., Eden Prairie, Minnesota
[3] Institute für Angewandte und Physikalische Chemie, Universität Bremen, Bremen, Germany
[4] Micromass Inc. Beverly, Mass.

ABSTRACT

We review here the recent characterization of vacuum deposited monolayer and multilayer thin films of two different perylenetetracarboxylic-dianhydride-bisimides (C_n-PTCDI; n =4,5), quinacridone, and two new bis-(N-alkyl)-quinacridone dyes (DIQA and DEHQA) on single crystal alkali halides using a combination of *in situ* luminescence spectroscopies and *ex situ* tapping mode AFM. Flat lying monolayer structures are indicated for PTCDA on the (100) faces of NaCl, KCl and KBr, for C4-PTCDI on KCl, for C5-PTCDI on both KCl and KBr and for DIQA on both KCl and KBr. Coherent thin films, exhibiting layer-by-layer growth can be achieved for PTCDA on all substrates, for C4-PTCDI on KCl and for DIQA on both KBr and KCl. Both C4-PTCDI and DIQA appear to fulfill the requirements for dyes which exhibit layered growth with the molecular plane inclined at steep angles to the surface normal.

INTRODUCTION

Charge transport in ordered organic thin films, as with their optical properties, is likely to be highly anisotropic with respect to the plane of the molecule. Vacuum deposition of crystalline organic materials can provide for formation of thin films in which these anisotropies may be studied [1,2], and is attractive for the creation of large scale device-quality thin films [3-6]. Ultimately control of molecular orientation may be an essential component of the processing of these materials in several emerging technologies. For organic field effect transistor (OFET) materials the preferred direction of transport in the organic thin film is parallel to the substrate plane, for photovoltaic (PV) and organic light emitting diode (OLED) applications the preferred direction of transport is perpendicular to this plane [3-8]. Charge carrier mobilities are increased in well ordered crystalline organic materials [2] but controlling dye orientation to take advantage of the inherent anisotropies in conductivities and mobilities has been challenging [8,9].

Control of the packing architecture of the dye layers can be dependent upon the structures developed in the first monolayer, and the way in which those structures are sustained or altered in multilayer versions of these thin films. Commensurate or coincident surface lattices are obtained for a wide variety of organic dyes on substrates ranging from single crystal metals, to surface-passivated silicon, layered semiconductors, HOPG, and alkali metal halides [9-15]. In many systems monolayer structures consist of "flat-lying" molecules where the plane of the aromatic dye is parallel, or nearly parallel, to the surface plane. Multilayers of such materials exhibiting true layer-by-layer growth only arise from dyes which form layered structures in the bulk, where the layer plane is parallel to the plane of the dye. Examples of such materials include the trivalent and tetravalent metal phthalocyanines, and the perylene tetracarboxylic-dianhydride, PTCDA (Figure 1) [11-15]. It appears from recent studies that layered bulk structures can also be obtained for certain other dyes where large angles exist between the plane of the dye and the layer plane [1,16]. The anticipated dependence of both optical and electrical properties on the orientation of the molecular plane of these dyes, with respect to both the substrate plane and the plane of contacting electrodes, motivates our interest in understanding the way in which layered growth is initiated in such dyes, starting with flat-lying monolayers.

In this paper we discuss the monolayer and multilayer packing architectures for *a)* PTCDA and related butyl (C4) and pentyl (C5) bisimides (C4-PTCDI and C5-PTCDI) on single crystal alkali halides, and *b)* a new bis-(N-iso-amyl)-quinacridone (DIQA), where comparisons are made relative to its parent molecule (QAD), and bis-(N-ethylhexyl)-quinacridone (DEHQA). The X-ray crystal structures for C4-PTCDI and C5-PTCDI have been published [25]. C4-PTCDI has a monoclinic structure, two molecules per unit cell, a = 0.47 nm b=2.82 nm, c=0.94 nm and β = 110.86°. With only one carbon difference in each imide chain, C5-PTCDI nevertheless forms a triclinic bulk structure, one molecule per unit cell, a = 0.475 nm, b = 0.848 nm, c = 1.63 nm, α = 86.88°, β = 83.50°, γ = 83.69°. These differences in crystal structure translate into quite different film morphologies for each dye.

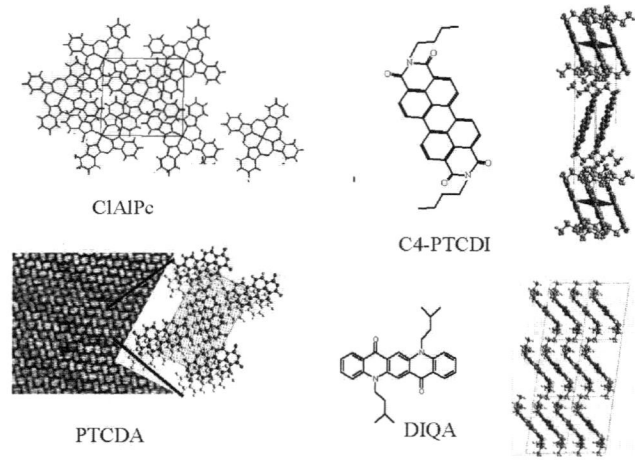

ClAlPc

C4-PTCDI

PTCDA

DIQA

Figure 1 *Monolayer and multilayer packing structures for ClAlPc, PTCDA, C4-PTCDI, and DIQA. The first and second layers of the ClAlPc structure are shown, as found on surfaces such as Cu(100), SnS₂(0001) and MoS₂(0001), and KCl (100) [12-15]. The packing structure for the first two layers of PTCDA is also shown, along with an STM image of a single monolayer of this material on Au(111) [16]. In contrast, the unit cell for C4-PTCDI, oriented with the b-axis vertical, and DIQA, with the a-axis vertical, provides for layered growth of these materials with the plane of the molecule not parallel to the substrate plane [25,27].*

PTCDA, C4-PTCDI and C5-PTCDI appear to form flat-lying monolayers on various alkali halide (100) surfaces, due to their strong interactions between the aromatic dye and the substrate. Likewise DIQA appears to form such monolayer structures on both KCl and KBr surfaces. C4-PTCDI and DIQA appear to restructure easily to their layered bulk forms following completion of the first monolayer, and evidence for the onset of layer-by-layer growth is seen in AFM images of 2-3 monolayer coverage films. C5-PTCDI does not achieve this layered growth, probably because of fundamental differences in its bulk structure relative to these other two dyes. Luminescence spectroscopies for all of these dyes show the transition from monomer-like to aggregated forms in the first 2-3 monolayers, with intermediate (Y-type) excimer species observed at the onset of multilayer film growth.

EXPERIMENTAL

The synthesis of C4-PTCDI, C5-PTCDI and DIQA has been discussed elsewhere [13,17]. The vacuum deposition chambers, luminescence spectrometers and AFM techniques used in these studies have been described previously [1,13]. Dyes were deposited at rates of less than 1.0 monolayer per minute on the (100) surfaces of freshly cleaved KCl, KBr and NaCl, substrates which allowed for the spectroscopic characterization (both visible and IR) of the deposited films [13]. Laser excitation was used (514 nm, 488 nm) from a 10 mW argon ion laser source, coupled into the vacuum deposition chamber. Light was collected at 90° to the incident beam, focused onto a J-Y H10 scanning monochromator, equipped with a PMT detector. Tapping mode AFM studies were conducted in air using a Nanoscope III (Digital Instruments), using oxide-sharpened silicon nitride tips.

Alkali halide crystals were cleaved using a rapid striking blow with a razor blade and immediately placed in vacuum. No annealing process was used. High temperature annealing and slow "push" cleave methods were found to increase the step density which is not desirable for studies of layer-by-layer growth [13]. Growth of many of the crystalline dyes is strongly affected by nucleation at step edges, therefore some dependence is seen in the motifs of crystal growth in the dyes discussed here.

RESULTS AND DISCUSSION

PERYLENETETRACARBOXYLICDIANHYDRIDE-BISIMIDE THIN FILMS

The perylenetetracarboxylicbisimide dyes explored to date (C4-PTCDI and C5-PTCDI) and PTCDA appear to form flat-lying monolayer structures on all alkali halide single crystal surfaces [18]. Varying degrees of interaction are possible as evidenced by differences in the luminescence response of these low coverage films [1,13,19]. Figure 2 shows the position of the luminescence band for dilute, monomeric, solutions of both of the bisimides, and the progression in energies of these luminescence bands for monolayer structures as the apparent strength of the interaction with the substrate material increases. Table 1 summarizes the positions of the 0-0 and 0-1 luminescence bands for this series of dyes in $CHCl_3$ solutions and on (100) surfaces of NaCl, KBr, and KCl, and on glass. For PTCDA the luminescence bands of the monolayer film on all three alkali halide surfaces were blue-shifted by more than 20 nm from that seen for dilute PTCDA solutions due to the interaction of the π-conjugated system with the substrate. Such monomer-like luminescence bands cannot be obtained on glass surfaces, except at temperatures near 10°K, another indication of the strength of interaction between PTCDA and these alkali halide surfaces [1,13,19]. For the C4-PTCDI system, only deposits formed on the KCl (100) surfaces showed this blue-shift in luminescence response for the monomeric species, suggesting that the interaction with both NaCl and KBr surfaces was significantly weaker than on KCl. The C5-PTCDI system showed these large blue-shifts in luminescence response on both KBr and KCl surfaces.

It has been proposed that the dominant interaction determining the molecular orientation of PTCDA on alkali halide surfaces is the dipole-induced-dipole interaction between the carbonyl oxygens of the PTCDA molecule and the surface cations [18]. The flat-lying PTCDA core on the NaCl (100) surface can be arranged with the long axis of the molecule in the <010> direction so that all four of the carbonyl oxygens are above surface cations with less than 5% lattice mismatch. On KCl and KBr (100) surfaces, only opposite pairs of carbonyl bonds can lie above surface cations, if the molecule is rotated so that these carbonyl oxygens lie along

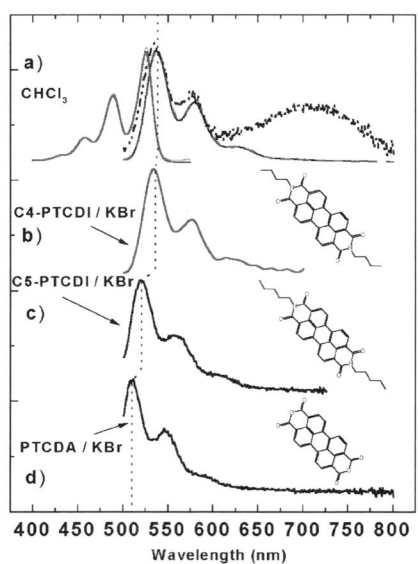

Figure 2 a) dotted line: Absorbance and luminescence spectra for PTCDA in CHCl₃, (luminescence spectrum shows aggregated characteristics, broad excimer luminescence band at ca. 710 nm), solid lines: C4-PTCDI and C5-PTCDI ca. 10^{-5}M in CHCl₃,; (b – d) luminescence spectra for monolayer coverages of C4-PTCDI, C5-PTCDI and PTCDA on the KBr (100) surface, showing the spectral shifts which result with varying degrees of interaction between the dye and the substrate [13].

Table 1 *Position in nm (and thousands of cm^{-1}) of 0-0 and 0-1 perylenetetracarboxylic-dianhydride and bisimide dye emission peaks in CHCl$_3$ and on alkali halide and glass surfaces.*

		CHCl$_3$	NaCl	KBr	KCl	Glass
C5-PTCDI	0-0	537 (18.6)	537 (18.6)	520 (19.2)	520 (19.2)	543 (18.4)
	0-1	581 (17.2)	580 (17.2)	565 (17.7)	565 (17.7)	590 (16.9)
C4-PTCDI	0-0	537 (18.6)	536 (18.6)	534 (18.7)	525 (19.0)	548 (18.2)
	0-1	581 (17.2)	581 (17.2)	580 (17.2)	564 (17.7)	590 (16.9)
PTCDA	0-0	537 (18.6)	517 (19.3)	509 (19.6)	516 (19.3)	n/a
	0-1	581 (17.2)	554 (18.1)	547(18.3)	550 (18.2)	

the <010> direction, with good lattice matching in the case of KCl (less than 3% mismatch) and less good in the case of KBr (ca. 8% mismatch) [13]. The C$_n$-PTCDI systems may create similar monolayer structures, but with increasing alkyl chain length would be constrained to tilt the alkyl chains out of the substrate plane. The luminescence data shown here suggest that C4-PTCDI on KCl and C5-PTCDI on both KCl and KBr surfaces start out in flat-lying, monomer-like architectures, and that formation of their stable bulk structures in a thin film requires a significant transition to different orientations of the molecular plane with respect to the substrate.

Figure 3a shows a representative tapping mode AFM image (5 x 5 micron) of a submonolayer deposit of C5-PTCDI on the KBr (100) surface. Coherent domains of the dye are seen on the terraces of the KBr surface, with x and y dimensions approaching 1 micron, and with a thickness of ca. 0.5 nm, consistent with a nearly flat-lying deposit of this perylene dye.

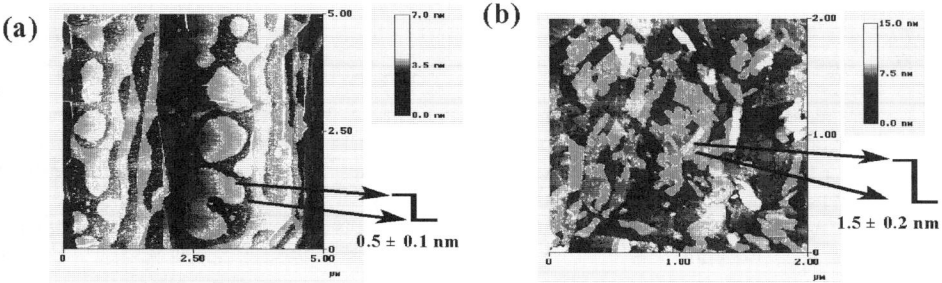

Figure 3 *Tapping mode AFM images for (a) a monolayer deposit of C5-PTCDI on KBr (100) – the crystalline regions of the dye have an average step height of 0.5 nm, and (b) a ca. 2-3 monolayer deposit of C4-PTCDI deposit on KCl (100).*

Nucleation of this dye appears to occur preferentially along step edges on the KBr surface, and the monolayer deposits do not appear to possess additional crystallinity or orientation with respect to the substrate, on this distance scale. Subsequent deposition of C5-PTCDI on these substrates does not lead to layer-by-layer growth on the distance scales seen for C4-PTCDI (Figure 3b, see discussion below) [13]. Instead, wall-like, ca. 1 micron length crystallites are seen growing on top of this first monolayer, suggesting weak interactions between layers, and no tendency for layered growth. Comparable results were obtained for the growth of C5-PTCDI on KCl (100).

Figure 4a shows normalized luminescence spectra for the successive deposition of submonolayer (0.05 ML) to multilayer (4.0 ML) deposits of C5-PTCDI on KBr (100). Comparable data were obtained for the growth of this molecule on KCl. Their general appearance is similar to that recently reported for ultrathin films of PTCDA and C4-PTCDI on various alkali halide surfaces [1,13]. At low coverages the spectrum is dominated by the luminescence of monomer-like species as discussed above. If deposition is stopped at these coverages, the 0-0 peak at ca. 520 nm slowly decays and new spectral features arise, in the region near 560 nm, near 600-620 nm, and finally near 650-670 nm. These new emission

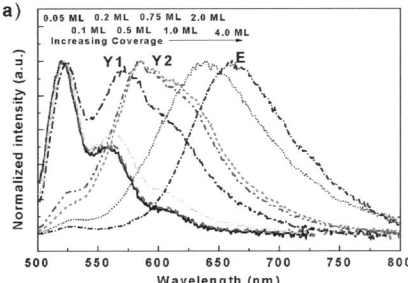 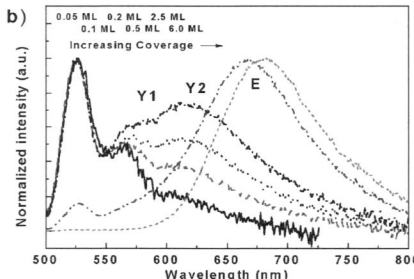

Figure 4 *Normalized luminescence spectra as a function of coverage for (a) C5-PTCDI on KBr (100); (b) C4-PTCDI on KCl (100).*

features have been assigned to different forms of excimers, the Y1, Y2 and E species respectively [19-22]. For C5-PTCDI on the KBr (100) surface as dye coverage is increased a distinct Y1 emission feature is seen, which then diminishes in relative intensity as the Y2 and E bands appear. The Y1 and Y2 spectral features appear to arise from the first formed (possibly dimeric) aggregates on these alkali halide surfaces and represent intermediate (strained) phases of crystal growth, on the way to achieving a stable bulk crystalline phase (the E-excimer form).

Figure 3b shows a typical tapping mode AFM image for a ca. 3 monolayer deposit of C4-PTCDI on a KCl (100) surface with sizeable terraces. Over regions of up to 0.3 microns this dye grows in a coherent layer-by-layer deposit, with the step height between the upper layers of dye equal to ca. 1.5 nm, the height expected from the b-axis dimension of the C4-PTCDI unit cell if the (010) layer planes in the crystal are parallel to the substrate plane (2.8 nm for two layers, as shown in Figure 1) [25]. Earlier AFM studies of lower coverages of C4-PTCDI on highly faceted KCl surfaces showed that the first deposited layers were ca. 0.5 nm thick, followed by a layer ca. 0.9 nm thickness. The combined interpretation of these studies suggests that the transition to the bulk structure occurs in stages. Interpretation of these transitions is complicated by step density differences in these surfaces, since edge sites are known to nucleate crystal growth and clearly help direct which crystal motifs are dominant. The intermediate layer may expose a different crystal face or may consist of a less defined structure existing only as a transition between the flat lying structure and the bulk crystal structure, perhaps stabilized by step edges in the salt surface. Luminescence spectral responses for submonolayer to multilayer thin films of C4-PTCDI on KCl (100) (with large terrace widths) are shown in Figure 4b. Relative to the deposition of C5-PTCDI on KBr, there is a more rapid transition from the monomer-like spectral features, through the Y2 excimer species, to the final E excimer form, at coverages of ca. 2 monolayers. This earlier transition in spectral response with increasing dye coverage suggests the easier formation of a stable aggregate for C4-PTCDI versus C5-PTCDI.

The addition of one extra carbon to the alkyl chain of C4-PTCDI to produce C5-PTCDI therefore leads to substantially different thin film morphologies, which probably originates in the differences in bulk structure of these two dyes. Figure 5 shows two views of the bulk structure for C5-PTCDI [25]. There is some overlap of the alkyl chains within the nearly clear (001) plane, a feature which is absent in the C4-PTCDI bulk structure. The fact that comparable coverages of C5-PTCDI on the same substrate gave a much less well ordered deposit is consistent with the difficulty in forming a clear layer plane parallel to the substrate plane. The extended transition from monomer-like and Y2-excimer spectral features to the final E-type excimer luminescence band is also consistent with the increased difficulty in forming a stable bulk structure in the thin film, versus the behavior seen for PTCDA and C4-PTCDI.

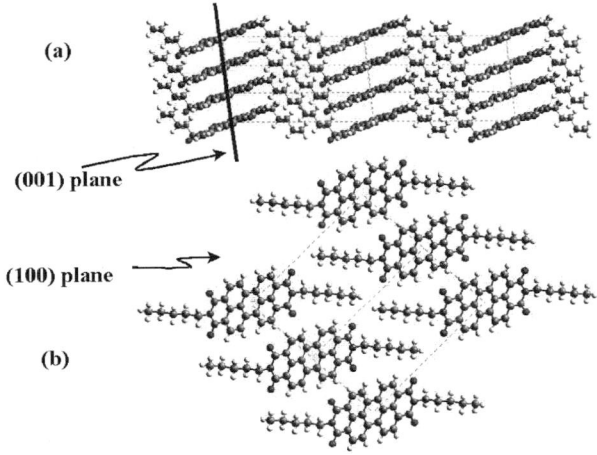

(a)

(001) plane

(100) plane

(b)

Figure 5 *Schematic views of the crystal structure for C5-PTCDI [25], showing the (001) and (100) planes. Overlap of adjacent hydrocarbon tails seems to preclude layered growth of this material.*

MODIFIED QUINACRIDONE DYES

Figure 6 shows the molecular structures of quinacridone (QAD), bis-(N-isoamyl)-quinacridone (DIQA), and bis-(N-ethylhexyl)-quinacridone (DEHQA), and AFM images for ca. 3-4 monolayer deposits of these dyes on either KCl or KBr (100), reflecting the extremes in film growth seen for these molecules. The differences in thin film morphology are reflective of the differences in packing structure brought about by the addition of side chains at the nitrogen-

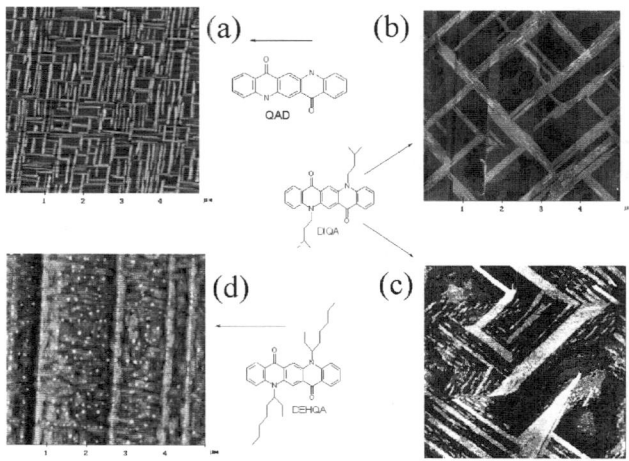

Figure 6 *Molecular structures of QAD, DIQA and DEHQA, and tapping mode AFM images(5 x 5 micron) of ca. 2-3 monolayer coverages of a) QAD on KCl (100), b) DIQA on KCl (100), c) DIQA on KBr, d) DEHQA on KCl.*

position on the quinacridone ring. The absorbance and luminescence spectra for these three dyes at approximately the same coverages on the KCl (100) surface are shown in Figure 7. A summary of the bulk single crystal structures of these dyes, obtained for QAD from recent literature [26] and for DIQA and DEHQA from recently completed single crystal structure determinations [27], is shown in Figure 8.

The parent dye, quinacridone (QAD), has a bulk structure in the γ-phase which suggests that layering of this material might occur [26]. Quinacridone (γ phase) is monoclinic, has two molecules per unit cell, with a = 1.38 nm, b = 0.39 nm, c =1.34 nm, β = 100.33° [26]. On all substrates studied to date, while there may be flat-lying, well ordered monolayers formed, the transition to a bulk structure is abrupt and is accompanied by a dewetting of the surface, starting with the deposition of the second layer, and does not lead to coherent, conformal films. As can be seen in Figure 6a, multilayer deposits of QAD on KCl (100) possess a specific alignment with the main crystallographic axes of the substrate, but prefer to form thin, "wall-like" motifs on the micron scale. The absorbance spectra of these films, shown in Figure 7, are red-shifted from those seen for dilute solutions of the monomer species and posses a progression of absorbance peaks to lower wavelength, suggestive of multiple aggregate forms of the dye [24]. The

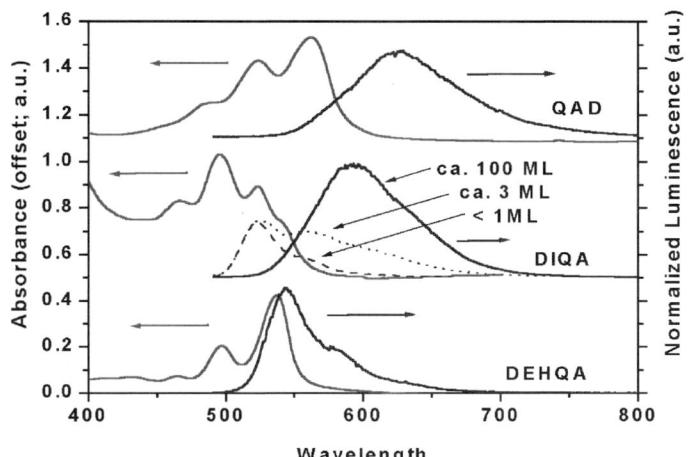

Figure 7 *Absorbance (obtained ex situ, for multilayer films, in transmission mode) and luminescence spectra for QAD, DIQA and DEHQA thin films on KCl (100). For DIQA the luminescence response for ca. 0.5 ML, 3 ML and 100 ML (normalized responses) are shown.*

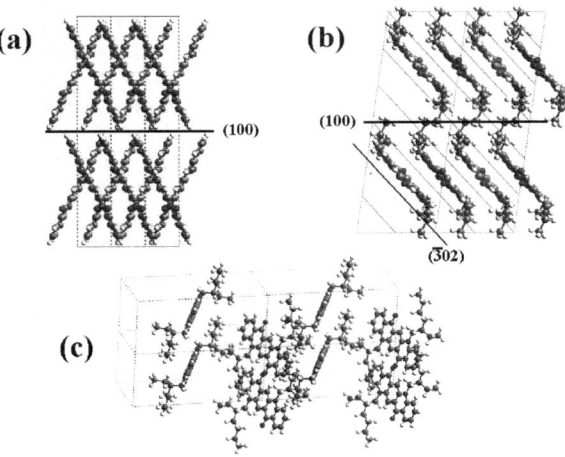

Figure 8 *Views from the bulk crystal structures for (a) quinacridone (the solid line is parallel to the (100) plane); (b) DIQA (the solid line is parallel to the 100 plane, and the diagonal lines are parallel to the −302 plane); (c) DEHQA (no clear layer plane exists for this structure).*

luminescence spectra of these films, even at low coverages of QAD, suggest an excimer (E-type) luminescence response, consistent with a strongly aggregated dye.

The addition of iso-amyl groups at the two nitrogen positions produces the dye DIQA with a bulk structure (Figure 8) possessing layer planes between adjacent hydrocarbon tail regions (along the (100) plane). The crystal structure for DIQA is monoclinic with two molecules per unit cell; a = 1.26 nm, b = 1.13 nm, c=0.86 nm, β = 98.08° [27]. The AFM data for this material on KCl and KBr surfaces (Figure 6) show evidence for layered growth. Deposition of the first monolayer (step height = ca. 0.5 nm) is followed with second layer step heights of ca. 1.0 nm (on KCl) and ca. 1.3 to 1.5 nm (on KBr). This last step height is close to the a-axis length of the unit cell, and provides for arrangement of the molecules as shown in Figure 8b, with the (100) plane parallel to the substrate plane.

It is interesting to note for both substrates that these layered structures have a clear orientation with respect to their major crystallographic axes. Both vertical and horizontal step edges in the alkali halide crystal are clearly visible in Figure 6b, identifiable by abrupt, sharply defined changes in shading in the image of the KCl (100) plane. The upper layers of DIQA all intersect step-edges, and appear to grow out from this edge into the monolayer-covered terrace region, at 45° angles. The implications of this growth mode are that registry with the first deposited layer is important, and that step-edges on these substrates help to initiate growth and define the crystal motifs of the deposited material. Molecular resolution AFM images have been elusive using these substrates, however, studies are underway to verify this type of growth pattern on substrates which support higher resolution AFM imaging, to better understand these film growth processes.

The absorbance spectra (Figure 7) of multilayer films of DIQA are consistent with the complex aggregate architecture, where the transition dipoles in adjacent dye molecules are inclined at compound angles [27]. Examination of the single crystal structure of this material shows that adjacent quinacridone units overlap only over ca. one phenyl unit, with a distance of closest approach of ca. 0.35 nm [26]. The luminescence spectra of DIQA, as a function of coverage on both KCl and KBr surfaces, show a clear transition from monomer-like species at less than one monolayer coverage, and the systematic transition to an E-type excimer of this species starting as coverage is increased above one monolayer.

DEHQA forms a stable bulk crystal, but the presence of the branched, chiral ethylhexyl side chains at the two nitrogen positions keeps adjacent quinacridone rings from cofacial aggregation, with no clear layer planes indicated (Figure 8c). The crystal structure for DEHQA is monoclinic with two molecules per unit cell; a = 0.78 nm, b = 1.09 nm, c=1.85 nm, β = 91.20° [27]. Interestingly, both the absorbance and luminescence spectra of multilayer films of DEHQA on KCl (100) indicate a monomer-like form of this dye, with a clear vibronic progression in the absorbance and luminescence spectra only slightly shifted from the peak values seen for this molecule in solution. That these absorbance and luminescence spectral features were obtained at all coverages of DEHQA confirms that exciton splitting effects are minimized in this molecule. AFM images such as those shown in Figure 6 show an overall orientation of the crystalline thin films on the micron scale with respect to the major crystallographic axes of the substrate, but no indication of layer-by-layer growth and a much more textured surface overall. For applications requiring high concentrations of this dye in an electroluminescent matrix, quenching of the luminescence response of this dye at high concentrations is much less likely than for either DIQA, QAD, or the bis-(N-methyl)- derivative of quinacridone [17], but it is clearly a less acceptable candidate for growth of thin films with clearly defined layer planes.

CONCLUSIONS

Strongly luminescent dyes like C4-PTCDI and DIQA lend themselves to characterization of their monolayer and multilayer growth patterns using a combination of *in situ* spectroscopies and *ex situ* AFM. The strength of interaction with the alkali halide single crystal surfaces appears to force flat-lying monolayer structures regardless of the bulk crystal structures for these dyes; this is a condition which is likely to persist for all planar substrates, perhaps even those where interaction with the dye is purely van der Waals in nature. This is an issue which is under active exploration for metal substrates covered with close packed self-assembled monolayers where the degree of interaction with the first monolayer can be more closely controlled.

Following the formation of this first monolayer, the formation of stable layered multilayers appears to be quite variable, and has only been conclusively seen for the C4-PTCDI and DIQA systems where layer planes can be found in the bulk structures. It appears that future studies of the luminescence response of dye layers undergoing structural transformations, from flat-lying monolayer to tilted, layered structures, can anticipate characteristic bands appearing in the luminescence spectra which correspond to intermediate, "strained" polymorphic forms of the aggregated dyes. The coverages at which these intermediate luminescence bands are seen may be correlated with the ease with which the final layered structures are formed. The degree to which these layered, conformal thin films can be produced in device (e.g. OFET) geometries, and the relationship between ordering of these dyes and their electrical properties, is currently under exploration.

ACKNOWLEDGEMENTS

This research was supported in part by grants from the National Science Foundation (Chemistry and International Programs), by the Center for Advanced Multifunctional Polymers and Molecular Assemblies (CAMP; an ONR-MURI program), and by the Materials Characterization Program, State of Arizona.

REFERENCES

1. D. Schlettwein, A. Back, B. Schilling, T. Fritz, N.R. Armstrong, *Chemistry of Materials*, **10**, 601 (1998).
2. A. Ioannidis, J.-P. Dodelet, *J. Phys. Chem. B*, **101**, 5100 (1997).
3. H. Klauk, T.N. Jackson, *Solid State Tech.*, **43**, 63 (2000).
4. Z. Bao, *Adv. Mater.*, **12**, 227, (2000).
5. Z. Bao, J.A. Rogers, H.E. Katz, *J. Mater. Chem.*, **9**, 1895 (1999).
6. C. Tang, S.A. VanSlyke, *Appl. Phys. Lett.*, **51**, 913 (1987).
7. J.J.M. Halls, J. Cornil, D.A. dos Santos, R. Silbey, D.H. Hwang, A.B. Holmes, J.-L. Bredas, R.H. Friend, *Phys. Rev. B Cond. Matt.* **60**, 5721 (1999).
8. P. Smolenyak, R. Peterson, K. Nebesny, M. Törker, D.F. O'Brien, N.R. Armstrong, *J. Amer. Chem. Soc.* **121**, 8628 [1999].
9. Šilerová, R; Kalvoda, L.; Neher, D.; Ferencz, A.; Wu, J.; Wegner, G. *Chem. Mater.*, **1998**, *10*, 2284.

10. J.A. Last, D.E. Hooks, A.C. Hillier, M.D. Ward *J. Phys. Chem. B.*, **103**, 6723 (1999).
11. A.C. Hillier, M.D. Ward, *Phys. Rev. B, Cond. Matt.*, **54**, 14037 (1996).
12. A. Schmidt, L.-K. Chau, A. Back, N.R. Armstrong, in *"Phthalocyanines,"* C. Leznoff, A.P.B. Lever, eds., VCH Publications, Volume **4**, 1996, pp. 307- 341, and references therein.
13. A. Schmidt, R. Schlaf, D. Louder, L.-K. Chau, S.-Y. Chen, T. Fritz, M. F. Lawrence, B. A. Parkinson, N.R. Armstrong, *Chem. Mater.*, **7**, 2127 (1995).
14. A. Back, Ph.D. Dissertation, University of Arizona, 1999.
15. I. Chizhov, G. Scoles, A. Kahn, *Langmuir*, (in-press).
16. T. Schmitz-Hubsch, T. Fritz, F. Sellam, R. Staub, K. Leo, *Phys. Rev. B. Condensed Matter*, **55**, 7972 (1997).
17. A. Tsuchida, S. Hayashi, G. Schnurpfeil. A. Ashida, D. Wörhle, H. Yanagi, *Chem. Funct. Dyes*, **2**, 775 (1993).
18. S.E. Shaheen, Y. Kawabe, J.-F. Wang, J.D. Anderson, E.A. Mash, P.A. Lee, N.R. Armstrong, B. Kippelen, N. Peyghambarian, *J. Appl. Phys.*, **85**, 7939 (1999).
19. M. Möbus, N. Karl, T. Kobayashi, *J. Cryst. Growth*, **116**, 495 (1992).
20. U Gómez, M. Leonhardt, H. Port, H. C. Wolf, *Chem. Phys. Lett.*, **268**, 1 (1997).
21. S. Akimoto, A. Ohmori, I. Yamazaki, *J. Phys. Chem. B*, **101**, 3753 (1997).
22. M. I. Sluch, A. G. Vitukhnovsky, M. C. Petty, *Thin Solid Films*, **284-285**, 622 (1996).
23. J. Mahrt, F. Willig, W. Storck, D. Weiss, R. Kietzmann, K. Schwarzburg, B. Tufts, B. Trösken, *J. Phys. Chem.* **98**, 1888 (1994).
24. U. Keller, K. Mullen, S. DeFeyter, F.C. DeSchryver, *Adv. Mat.*, **8**, 490 (1996).
25. E. Hädicke, F. Graser, *Acta Cryst.* **C42**, 189 and 195 (1986), and G. Klebe, F. Graser, E. Hädicke, J. Berndt, *Acta Cryst.*, **B45**, 69 (1989), and references therein.
26. G.D. Potts, W. Jones, J.F. Bullock, S.J. Andrews, S.J. Magnon, *J. Chem. Soc. Chem. Comm.*, **1994**, 2565; and G. Lincke, H.-U. Finzel, *Cryst. Res. Technol.*, **31**, 441, (1996).
27. D. Alloway, J.-F. Wang, B. Schilling, M. Carducci, N.R. Armstrong, manuscript in preparation.
28. M. Kasha, in *"Spectroscopy of the Excited State,"* B.D. Bartolo, ed., Plenum Press, New York, 1976, pp 337-363, and references therein.

Molecular multilayers: structure and templating effects

Sandrine Heutz, Sallie M. Bayliss, Rudi Cloots [1], Ruth L. Middleton, Garry Rumbles and Tim S. Jones
Department of Chemistry and Centre for Electronic Materials and Devices,
Imperial College, London SW7 2AY, U. K.
[1] Laboratoire de Chimie Inorganique Structurale, Departement de Chimie B6,
Universite de Liege, B-4000 Liege, Belgium

ABSTRACT

Powder X-Ray diffraction (XRD) has been used to study multilayered structures grown by organic molecular beam deposition based on the molecular materials PTCDA and metal-free phthalocyanine (H_2Pc). Double layers of different polymorphic forms (α, β_1 and β_2) of H_2Pc indicate that the structure of the second layer is determined by the properties of the first layer. It is also shown that the first layer completely disrupts the crystallinity of the second layer in heterostructures containing PTCDA and H_2Pc. The implication is that a strong templating effect occurs during the growth of multilayer molecular thin film structures.

INTRODUCTION

The study of molecular materials has undergone a rapid development, since structures based on thin organic films offer significant potential for a variety of electronic and optoelectronic applications [1]. Archetypal molecular materials that have been intensively studied include the two conjugated planar molecules PTCDA and H_2Pc.

PTCDA crystallises in the $P2_1/c$ space group and the molecules lie parallel to the (102) plane. In this plane, the PTCDA molecules adopt a herringbone structure with an angle of about 45 ° between their long axes and the molecules arrange in columns [2]. Two polymorphs with slightly different lattice parameters have been identified in thin films formed by vapour deposition depending on the growth conditions and the type of substrate [3], but growth on glass and amorphous substrates only produces the α-polymorph. Free-base phtahlocyanine (H_2Pc) can also be grown as two different polymorphs [4]. The α phase is obtained by growth at room temperature, while annealing an α film, or high substrate temperature growth, produces the β phase. Both phases are monoclinic and characterised by a herringbone structure in which molecules stack along the **b**-axis. The α phase crystallises in the $C2/c$ space group [5], while the β form belongs to the $P2_1/a$ space group [6]. More recent studies have shown structural and morphological evidence that the β phase can be further differentiated as β_1 and β_2, for the annealed and the high substrate temperature growth modes respectively [7].

Electronic and optoelectronic device structures are usually based on two or more layers of molecular materials. It is thus important to know how the structure of the first layer will affect subsequent growth of different materials, or of an identical material deposited under different conditions. A number of studies have been made on periodic ultrathin multilayers, with particular emphasis on the electrical and optical properties of the heterojunction [8], or its structure [9], but the behaviour of thicker films which are generally used in device structures still needs to be assessed.

In this paper, we investigate the effect of the first layer on the structural properties of the second in a number of different systems. Double layers based on different polymorphs of H_2Pc are initially considered, and a strong templating effect is observed. The effect is then found to extend to heterostructures containing both PTCDA and α-H_2Pc.

EXPERIMENTAL

The films were grown by organic molecular beam deposition in a ultra-high vacuum chamber with a base pressure of ~ 2×10^{-9} torr. Commercial H_2Pc (SynTech, 99%) and PTCDA (Fluka, 98%) powders were outgassed for ~ 15 and 30 hours respectively prior to deposition onto glass substrates. The materials were sublimed by two separate miniature Knudsen effusion cells, held at a temperature of 340 °C for H_2Pc and 370 °C for PTCDA. This corresponds to a growth rate of ~ 5 Å s^{-1}, as determined by a quartz crystal microbalance (QCM) positioned near the substrate. The thickness of the samples is was 190 nm for the H_2Pc films and 140 nm for the PTCDA layers. The glass substrates were cut out of microscope slides (BHD super premium) and cleaned in a methanol sonic bath prior to transfer into the chamber.

The PTCDA films were grown at room temperature (α-phase), but three growth modes were used for the polymorphic H_2Pc double layers: (i) growth at room temperature (α phase), (ii) room temperature growth followed by annealing at 320 °C for 2 hours (β_1 phase) and (iii) high temperature growth at 320 °C (β_2). *Ex-situ* structural analysis was performed using a Siemens D5000 powder X-ray diffractometer.

RESULTS

H_2Pc double layers

A preliminary characterisation of the single layers is necessary before studying the multilayered structures. The different polymorphic forms of H_2Pc have been fully analysed elsewhere [7] but for the purpose of this paper we present a summary of the structural data. Powder XRD provides a clear differentiation between the different films studied, figure 1.

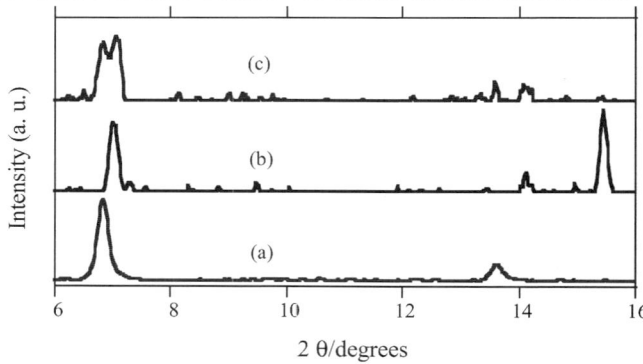

Figure 1 *Powder XRD 2θ scans for 190 nm H_2Pc films grown under different conditions: (a) α–phase, (b) β_1-phase and (c) β_2-phase.*

H_2Pc films grown at room temperature show two peaks at $2\theta = 6.8$ and $13.6\degree$, characteristic of diffraction from the (200) and (400) planes of the α phase (a). Annealing of the α film to produce a β_1 sample results in a diffraction scan with peaks at $2\theta = 7.0$, 14.1 and $15.5\degree$ (b). These correspond to diffraction from the (001), (002) and $(4\,\overline{1}\,\overline{3})$ planes of the β phase respectively. High substrate temperature growth (β_2) leads to a mixture of the patterns due to the α and β phase (c), but the peak at high angle seen for β_1 films is not present. Post-growth annealing of the β_2 film eliminates the diffraction peaks due to the α phase but does not induce the appearance of the $(4\,\overline{1}\,\overline{3})$ peak. The existence of the $(4\,\overline{1}\,\overline{3})$ peak depends therefore on the growth mode and is a characteristic of the β_1 phase.

Four different combinations of polymorphic double layer strcutures have been grown, as shown schematically in figure 2(a-d). A β_1 first layer can be followed by growth at room temperature (β_1/α) or high substrate temperature (β_1/β_2). When the first layer is of the β_2 type, subsequent growth, with and without annealing, leads to the β_2/β_1 and β_2/α combinations. The first layer must be of the β type since an initial α layer would be transformed into a β_1 film when high temperature is applied to the sample to grow the β second layer.

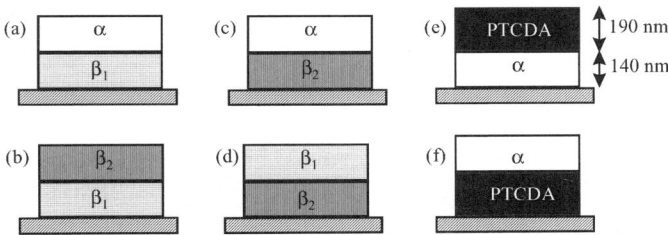

Figure 2 *Schematic of the combinations of double layers studied. The polymorphic H_2Pc double layers are (a) β_1/α, (b) β_1/β_2, (c) β_2/α and (d) β_2/β_1. Heterostructures based on both α-H_2Pc (abbreviated H_2Pc) and PTCDA are (e) $H_2Pc/PTCDA$ and (f) $PTCDA/H_2Pc$. The H_2Pc and PTCDA layers are 140 nm and 190 nm thick respectively.*

The XRD scans for the polymorphic double layers are shown in figure 3. The β_1/α sample (a) only shows the three diffraction peaks of the β_1 phase and there is no trace of the peaks due to the α phase. This indicates that the first β_1 layer has acted as a template for further growth at room temperature. A similar pattern is observed for the β_1/β_2 sample (b): the β_1 layer inhibits the growth of the α peaks accompanying the β_2 films and has again acted as a structural template. However, the $(4\,\overline{1}\,\overline{3})$ peak is not as intense as expected for a sample with β_1 characteristics. For a single layer (figure 1(b)), the $(4\,\overline{1}\,\overline{3})$ reflection is about as intense as the (001), while the (001)/$(4\,\overline{1}\,\overline{3})$ ratio is about three for the double layer. Therefore, the initial β_1 layer acts as an imperfect template in the β_1/β_2 structure.

Growth at room temperature onto the β_2 film (c) leads to a structure with mainly α characteristics. This is not surprising since the β_2 single layer has both α and β structural characteristics. The top layer preferentially matches the α crystallites contained in the β_2 layer.

Growth at room temperature onto the β_2 layer, followed by annealing, the β_2/β_1 double layer. As seen from figure 3(d), the α peaks accompanying the β_2 first layer disappear due to the post-growth anneal. The strong templating effect of the β_2 film is illustrated by the absence of the $(4\bar{1}\,3)$ peak, which is characteristic of a single β_1 layer.

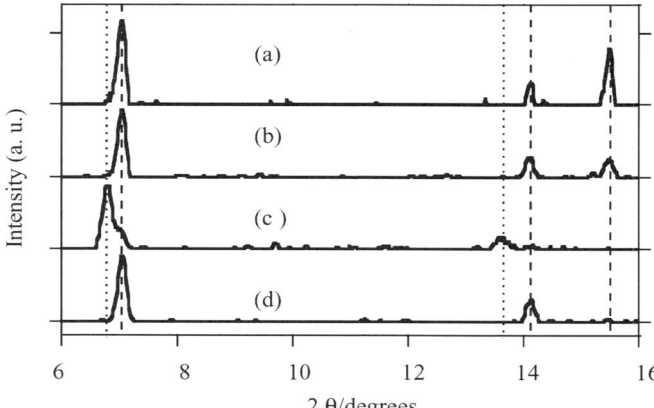

Figure 3 *X-ray diffraction scans of the polymorphic H_2Pc double layers: (a) β_1/α, (b) β_1/β_2, (c) β_2/α and (d) β_2/β_1. The vertical dashed lines represent peaks characteristic of the β-phase and the dotted lines those of the bulk α crystals.*

The β_2 layer inhibits the appearance of the $(4\bar{1}\,\bar{3})$ peak when a β_1 layer is deposited on top of it, while the β_1 layer appears unable to induce the formation of the $(4\bar{1}\,\bar{3})$ reflection in the β_1/β_2 structure. High temperature growth is a rapid process, and the β_2 layer grown on β_1 has little time to diffuse and adjust its crystal orientation to the first layer. However, when β_1 is grown on β_2, a pseudo α layer is initially formed, and slow diffusion allows it to match the structure of the first layer.

It is clear that the second layer in the H_2Pc-based structures mimics the structural behaviour of the first layer and that some type of homoepitaxial growth is occuring. More extensive studies on H_2Pc polymorphic double layers indicate that the first layer also acts as a morphological template for the growth of multilayer structures [10].

H$_2$Pc/PTCDA multilayers

In order to assess if the templating effect observed for polymorphic double layers also extends to heterostructures, double layers containing both H_2Pc and PTCDA were studied, see figure 2(e-f). Both films were grown at room temperature so that the H_2Pc film is of the α type. Figure 4 shows the diffraction scans for (a) α-H_2Pc and (b) PTCDA single layers, (c) H_2Pc/PTCDA, (d) PTCDA/H_2Pc and (e) PTCDA/H_2Pc/PTCDA/H_2Pc multilayers. The intensities of the diffraction scans have not been scaled and reflect the crystallinity of the two layers, allowing qualitative comparisons between the different samples.

PTCDA single layers grown on glass substrates are crystalline (b) and show a single diffraction peak at 27.6 °, corresponding to reflection on the (102) planes of the α-PTCDA [2] with an interplanar spacing of 3.22 Å. When PTCDA is grown on a H_2Pc film, only the peaks due to the H_2Pc first layer show in the diffractogram (c). The crystallography of PTCDA is completely disrupted and its characteristic peak at 27.6 ° has disappeared. H_2Pc has hence acted as a template on the subsequent growth of the PTCDA film. It should also be noted that the intensities of both α-H_2Pc reflections has increased with respect to the H_2Pc single layer.

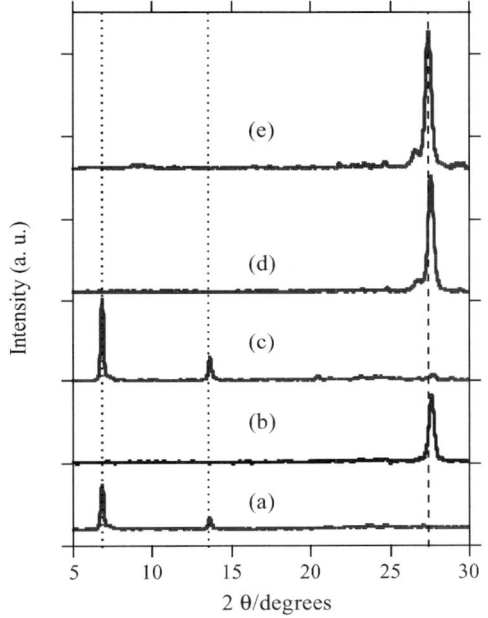

Figure 4 *XRD 2θ scans for reference single layers of (a) α-H_2Pc and (b) PTCDA and for the heterostructures: (c) H_2Pc/PTCDA, (d) PTCDA/H_2Pc and (e) PTCDA/H_2Pc/PTCDA/H_2Pc. The vertical dotted lines represent the positions of the reference peaks in the bulk H_2Pc α crystals, and the dashed line the one of PTCDA films.*

It is clear on the multilayer scans that the peaks that are representative of the top layers are not apparent.

An similar templating effect is observed when the deposition order is reversed, i.e. when H_2Pc is grown on a PTCDA first layer (d). Here, only the PTCDA diffraction signal is present and the H_2Pc contribution is absent. Only the peak that is characteristic of the first layer of material is observed and the PTCDA has templated the growth of second layer H_2Pc. Close inspection of the peak at ~ 27.6 ° reveals two unusual features. Firstly, the intensity of the peak in the double layer is increased with respect to the PTCDA single layer. Secondly, an additional shoulder grows at lower diffraction angles. After deconvolution of the peak the shoulder is found to be centred at 26.8 °, corresponding to an interplanar spacing d = 3.33 Å.

The templating effect extends through mulitlayered structures, as shown by the single diffraction peak of the PTCDA/H_2Pc/PTCDA/H_2Pc stack (e). The second layer, H_2Pc, is templated by the PTCDA and can not recover its crystal structure to template the next PTCDA layer, which then templates the last H_2Pc layer. A shoulder is again observed at the low angle side of the PTCDA diffraction peak.

The templating effect in the heterostructures is a complex issue. In contrast to the polymorphic H_2Pc double layers, the structural differences can not be attributed directly to epitaxial growth of the second layer. Despite the fact that both materials are polyaromatic planar molecules, the shape of the molecules and their crystal packing differs significantly. Even with these discrepancies, the increased intensity of the peaks characteristic of the first layer in the heterostructures suggests some type of epitaxy. The low angle shoulder at $2\theta = 26.8$ ° in the PTCDA/H_2Pc and multilayered samples could be due to a relaxation of the epitaxially grown H_2Pc to a larger interplanar spacing, closer to the 3.4 Å intermolecular spacing observed in H_2Pc bulk crystals.

Raman spectroscopy studies of these films suggest a top layer crystal packing similar to the unstrained single layer structure. It must be assumed that the top layer relaxes to its bulk structure after strained growth at the interface. The absence of a characteristic signal in the diffraction scan could be due to microcrystallinity. Alternatively, the crystallites may have different orientations, so that there is no diffraction plane parallel to the substrate in a suitable position for reflection. The templating effect in heterostructures seems to be a two-step process. Epitaxial growth at the interface induces either a small crystal size or a new orientation of the molecules on the substrate.

CONCLUSIONS

A strong templating effect has been observed for molecular films grown by organic molecular deposition. In the case of H_2Pc polymorphic double layers, it was shown that the second layer adopts the structure of the first layer independent of the growth conditions used. Heterostructures containing both H_2Pc and PTCDA layers display a similar effect and the structure of the top layer is drastically perturbed by growth on a different material. We believe that the second layer initially grows epitaxially on the bottom film. It subsequently relaxes into its bulk structure, either as microcrystallites or with the usual diffraction plane not parallel to the substrate. The templating effect has profound implications for molecular materials based devices, where multilayered structures are frequently used.

REFERENCES

1. S. R. Forrest, *Chem. Rev.*, **97**, 1793 (1997).
2. S. R. Forrest, M. L. Kaplan and P. H. Schmidt, *J. Appl. Phys.*, **55**, 1492 (1984).
3. M. Mobus, N. Karl and T. Kobayashi, *J. Cryst. Growth*, **116**, 495 (1992).
4. N. B. McKeown, *Phthalocyanine Materials*, (Cambridge University Press 1998), pp 41-47.
5. J. Janczac and R. Kubiac, *J. Alloys Comp.*, **190**, 121 (1992).
6. J. M. Robertson, *J. Chem. Soc.*, 1195 (1936).
7. S. M. Bayliss, S. Heutz, G. Rumbles and T. S. Jones, *Phys. Chem. Chem. Phys.*, **1**, 3673 (1999).
8. J. Danziger, J.-P. Dodelet, P. Lee, K. W. Nebesny and N. R. Armstrong, *Chem. Mater.*, **3**, 821 (1991).
9. N. Nanai, M. Yudasaka, Y. Ohki, S. Yoshimura, *Thin Solid Films*, **265**, 1 (1995).
10. S. M. Bayliss, S. Heutz, G. Rumbles and T. S. Jones, *Adv. Mater.*, **12**, 202 (2000).

Mat. Res. Soc. Symp. Vol. 620 © 2000 Materials Research Society

Surface morphology studies of phthalocyanine thin films:
mechanism of the α → β₁ phase transition

Sandrine Heutz, Sallie M. Bayliss, Garry Rumbles and Tim S. Jones
Centre for Electronic Materials and Devices and Department of Chemistry,
Imperial College, London SW7 2AY, United Kingdom

ABSTRACT

Free base phthalocyanine films have been grown on glass substrates by organic molecular beam deposition. In situ post-growth annealing of the samples leads to the $\alpha \rightarrow \beta_1$ transformation. Different transition states have been identified and their morphological properties studied by atomic force and optical microscopy. The transition occurs via a discrete number of nucleations and is preceded by an elongation of the α crystallites. The β_1 crystallites grow but are confined to domains of similar orientation. Increased thickness produces larger domains and better orientation, while only partial transformation occurs below 94 nm.

INTRODUCTION

Molecular thin films are an attractive alternative to more conventional inorganic materials for applications in electronics and optoelectronics [1]. The polyaromatic phthalocyanine films have a wide range of applications in solar cells [2], field emission transistors [3] and organic light emitting devices (OLEDs) [4]. Phthalocyanines exist as a variety of polymorphs and the most common phases can be grown by organic molecular beam deposition in a ultra-high vacuum (UHV) chamber [5]. The α phase of metal-free phthalocyanine (H_2Pc) can be obtained by growth at room temperature on weakly interacting substrates such as glass [6]. Annealing of an α film, or high temperature growth produce the β phase, which can be further differentiated as β_1 and β_2 [7]. The morphological, structural and spectroscopic properties of these three types of films have been fully assessed [7], but little is known about the mechanism of the transformation and the parameters affecting it.

In this paper, we study the $\alpha \rightarrow \beta_1$ transformation by varying the annealing time and the initial thickness of the film. In the early stages of the transformation, the α crystallites become elongated and the β_1 phase nucleates across the film and domains of slender crystallites of similar orientation gradually cover the surface of the entire film. Three regimes of behaviour have been identified depending on the thickness of the initial film, illustrating the strong influence of both crystal size and film continuity.

EXPERIMENTAL

The films were grown by organic molecular beam deposition in a UHV chamber with a base pressure of $\sim 2 \times 10^{-9}$ torr. The H_2Pc powder (SynTech, 99 %) was outgassed prior to growth and sublimed onto the substrate using a miniature Knudsen effusion cell held at ~ 330 °C. The growth rate was ~ 5 Å s^{-1} and the film thickness was calibrated using a quartz crystal microbalance positioned near the substrate.

Deposition was carried out onto glass substrates cut from microscope slides (BDH super premium) and cleaned thoroughly in a methanol sonic bath.

The $\alpha \rightarrow \beta_1$ transition was affected by *in-situ* annealing of samples grown at room temperature. Two series of samples were investigated. Firstly, 233 nm thick films were submitted to an annealing at 325 °C for various times, up to a maximum of 2 hours. Secondly, the thickness of the initial α film was varied for a given annealing time and temperature (325 °C for 2 hours). Five thicknesses were considered: 56, 66, 94, 233, and 481 nm. *Ex-situ* morphology analyses were performed using a Nomarski interference optical microscope (Olympus BH2-UMA) and an atomic force microscope (Burleigh instruments) operating in the tapping (non-contact) mode.

RESULTS AND DISCUSSION

The morphology of 233 nm films annealed for various times was studied by optical interference microscopy, figure 1. The reference α film with no post-anneal is completely smooth and only the imperfections of the glass substrate are visible (a). Distinctive features start to appear after 1 hour annealing, and irregular islands are seen on the smooth α-H_2Pc background (b). These islands are thought to be nucleation sites of the β_1 phase. The entire surface of the sample is covered by the islands after 1.25 hours annealing (c). No additional morphological changes were observed with longer annealing times.

a)   b)   c)

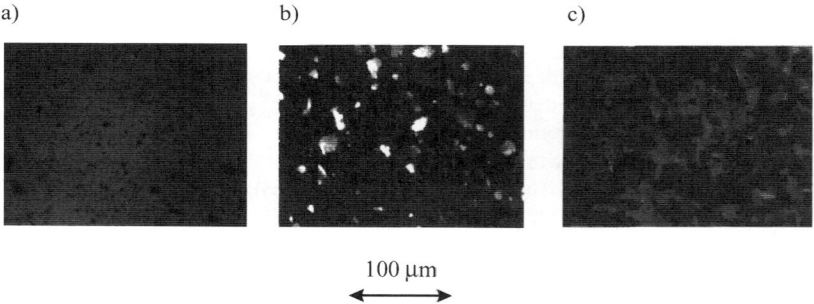

100 μm

◄——————►

Figure 1 *Nomarski micrographs of 233 nm thick H_2Pc films after annealing at 320 °C for: (a) 0, (b) 1.0 and (c) 1.25 hours.*

AFM was also used to assess the morphology of the films with increased annealing time, figure 2. The reference micrograph for the sample grown at room temperature shows the spherical crystallites typical of the α phase (a). After 0.75 hours, some elongation of these crystallites takes place, and the sphere diameter increases from ~ 0.1 μm to ~ 0.2 μm. A similar elongation prior to transformation was observed for CuPc on muscovite [8]. Elongation of the spheres becomes very visible after 1 hour annealing (b). A small proportion of this sample also shows slender elongated crystallites, parallel to each other over a short range, a morphology characteristic of the β_1 phase (c). These β_1 areas correspond to the poorly defined islands seen by optical microscopy and their size is ~ 50 μm^2.

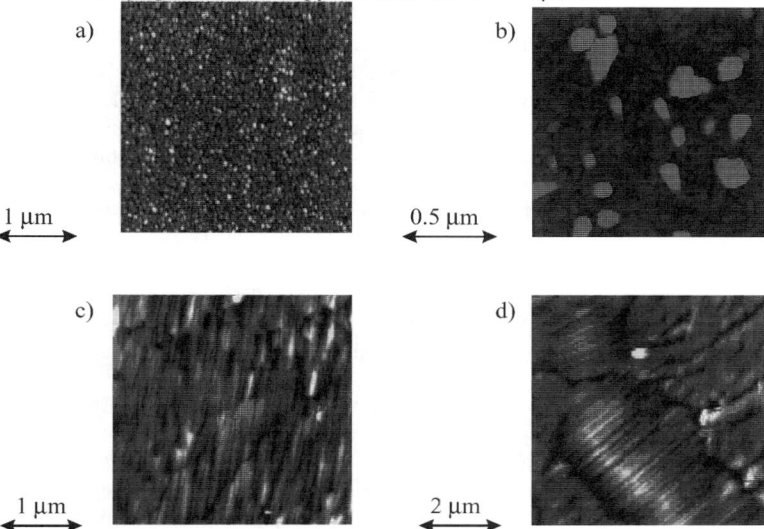

Figure 2 *AFM images of 233 nm thick H_2Pc films as a function of annealing time at 325 °C: (a) no anneal, (b) 1.0 h, α-phase, (c) 1.0 h, β_1-phase (minor component) and (d) 2.0 h.*

After 1.25 hours the surface is completely covered with the long crystallites of the β_1 phase, which is again consistent with the Nomarski analysis (figure 1(c)). XRD, electronic absorption and Raman spectroscopy results also confirm that there is an abrupt $\alpha \rightarrow \beta$ transition between 1.0 and 1.25 hours annealing [9]. Prolonged annealing induces the merging of the crystallites into smooth areas, while the remaining defined crystallites are arranged in columns (d). Boundaries between two areas with different crystallite orientation can also be seen, and large scale AFM images show that these correspond to the boundaries between the domains seen in the corresponding Nomarski micrographs. The domains seen by optical microscopy correspond to regions with similar crystalline orientation.

A quantitative size analysis has been performed on the AFM and Nomarski images of films submitted to various annealing times up to a maximum of 2 hours. The full details are published elsewhere [9]. The β_1 crystallites undergo a clear linear increase in size from 0.10 μm^2 (0.75 hours) to 0.52 μm^2 (2 hours). It is expected that the size of the crystallites will continue to increase until sublimation reduces the thickness to a level where gaps appear between the crystallites. A detailed analysis of the crystallite shape shows that both the α and β_1 crystallites become progressively more elongated, indicating growth along the **b**-axis. The domain size observed by optical microscopy follows a different trend: the island size increases until ~ 1.25-1.5 hours annealing, when it saturates at approximately 100 μm^2. Since the domains have been shown to be regions where the crystallites have different orientations, it is unlikely that the boundaries between these domains would break down to allow for the merging of the islands. The saturation is to be expected and each domain corresponds to a nucleation site of the β_1 phase. The number of nucleation sites is estimated at $1.2 \pm 0.3 \ \mu m^2$.

AFM and optical interference microscopy have also been used to study the influence of film thickness on the $\alpha \rightarrow \beta_1$ transformation, for a 2 hours annealing time at 325 °C. It was first noted that there is a film thickness constraint on the phase transition. It is not possible to effect the transformation on films which are thinner than 56 nm, either by growth at an elevated substrate temperature, or by annealing. This could either be due to a lack of material to keep a continuous film during the transition [10], or to the crystal size dependence of the relative stability of the two phases, since calculations have shown that the α-phase is more stable than the β phase for crystals containing less than $2x10^4$ molecules [11]. Assuming that the α crystallites are spherical, this corresponds to a crystal radius of 177 Å, or a diameter of 345 Å. This compares well with the observed film thickness of 56 nm, below which the transformation does not appear to occur and sublimation of the material takes place instead.

The initial film thickness has a dramatic influence on its morphology after annealing, as illustrated by the Nomarski micrographs in figure 3. For the thinnest film undergoing transformation (66 nm), the islands have a fern-like appearance and are separated by areas of untransformed α phase material (a). The surface becomes completely covered in domains for films thicker than 94 nm (b). It is also apparent that the domain size increases with film thickness (c), which may be due to a higher alignment of the crystallites in thicker films.

a)   b)   c)

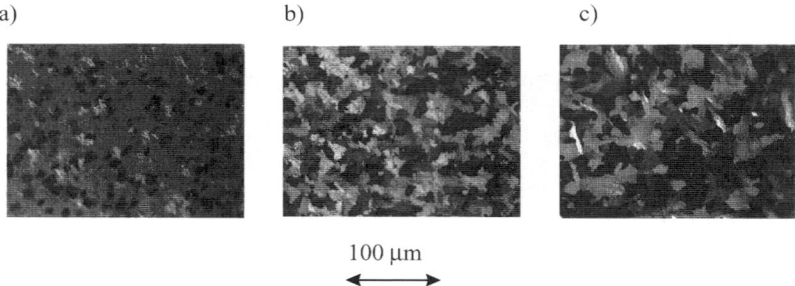

100 μm

Figure 3 *Nomarski micrographs of (a) 66 nm, (b) 94 nm and (c) 481 nm films after annealing at 320 °C for 2 hours.*

AFM analysis of the films at different thickness is presented in figure 4. The thinner films (< 66 nm) show two distinct regions having morphological characteristics of the β_1 and α phase respectively (a). The gaps between the two phases are consistent with Ashida's [10] observation that the $\alpha \rightarrow \beta$ transition is inhibited by the lack of material below a critical film thickness. For a 94 nm film (b), the surface is covered entirely in elongated crystallites, parallel to each other over a short range, so that many boundaries are present. The 233 nm film shown in figure 2(d) displays the same characteristics, with additional merged domains. The thickest film, figure 4(c) shows highly oriented β_1 crystallites and this confirms the previous observation that a higher thickness facilitates parallel orientation of the crystallites.

The absence of smooth merged domains can be attributed to a slow thermal energy transfer from the bottom heating source to the top layers of the film.

a)   b)   c)

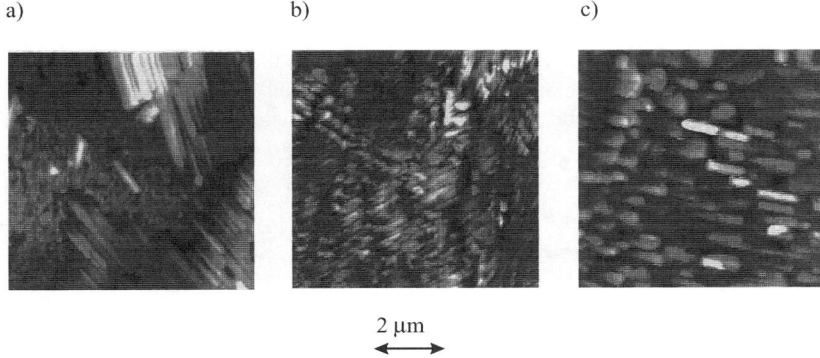

2 μm
◀━━━━▶

Figure 4 *AFM images of (a) 66 nm, (b) 94 nm and (c) 481 nm films annealed at 320 °C for 2 hours. Discrete domains of α and β₁ phase are observed for the thinnest film, where partial transformation occurs.*

The above observations indicate that there are three regimes of behaviour for films of different thickness. For films thinner than 56 nm, there is not enough material to allow any $\alpha \rightarrow \beta_1$ transformation. This is consistent with the suggestion that the α-phase is more stable than the β for small particles. For films between 56-94 nm, the transformation occurs but is not completed because there is not enough material on the surface to ensure a continuous film during the transition. Clearly, the β_1 phase grows at the expense of the α-phase directly in contact with it. The transformation occurs via discrete nucleations, since the remaining α areas coexist and do not intermix with completely transformed β_1 phase regions. Also, when the transformation is very close to the sublimation point, the regions of the film that have not been transformed are thin enough to be in the regime where the α phase is more stable than the β. For films above 94 nm, both the size of the α-phase crystallites and the thickness of the film (and hence its continuity) allow complete transformation of the film.

CONCLUSIONS

This study has elucidated the mechanism of the $\alpha \rightarrow \beta_1$ transformation obtained by annealing H_2Pc films. The spheres that are characteristic of the α phase undergo an elongation before slender, highly orientated β_1 crystallites nucleate at discrete points across the surface. Totally transformed films are completely covered in β_1 crystallites oriented over a short range in 100 μm^2 domains. When annealing is prolonged, the size of the domains saturates because of boundaries, while the individual crystallites grow along their **b**-axis and finally merge.

The thickness of the film also plays an important role and three regimes of behaviour have been distinguished. Thin films (< 56 nm) sublime without transformation, and partial transformation is observed for thicknesses between 56-94 nm. Increasing thickness produces completely transformed β_1 samples (> 94 nm) with increasing domain size.

ACKNOWLEDGEMENTS

SH thanks the Department of Chemistry, Imperial College for a postgraduate teaching assistantship. SMB is grateful to the Engineering and Physical Sciences Research Council (EPSRC), UK, for the provision of a Ph.D studentship. This work was supported by the EPSRC, UK.

REFERENCES

1. S. R. Forrest, *Chem. Rev.*, **97**, 1793 (1997).
2. C. W. Tang, *Appl. Phys. Lett.*, **48**, 183 (1986).
3. Z. Bao, A. J. Lovinger, and A. Dodabalapur, *Appl. Phys. Lett.*, **69**, 3066 (1996).
4. A. Van Slyke, C. H. Chen and C. W. Tang, *Appl. Phys. Lett,.* **69**, 2160 (1996).
5. N. B. McKeown, *Phthalocyanine Materials*, (Cambridge University Press 1998), pp 41-47.
6. S. Nespurek, H. Podlesak, and C. Hamann, *Thin Solid Films,* **249**, 230 (1994).
7. S. M. Bayliss, S. Heutz, G. Rumbles and T. S. Jones, *Phys. Chem. Chem. Phys.*, **1**, 3673 (1999).
8. M. Ashida, N. Uyeda and E. Suito, *J. Cryst. Growth,* **8**, 45 (1971).
9. S. Heutz, S. M. Bayliss, R. L. Middleton, G. Rumbles and T. S. Jones, *J. Phys. Chem. B*, submitted for publication.
10. M. Ashida, N. Uyeda and E. Suito, *Bull. Chem. Soc. Jpn.,* **39**, 2616 (1966).
11. F. Iwatsu, *J. Phys. Chem.,* **92**, 1678 (1988).

Biogenic and Biomimetic
Systems

Mat. Res. Soc. Symp. Vol. 620 © 2000 Materials Research Society

Templated Crystallization of Calcite on Patterned Self-Assembled Monolayers

Joanna Aizenberg
Bell Labs/Lucent Technologies
600 Mountain Ave.
Murray Hill, NJ 07974

ABSTRACT

Micropatterned self-assembled monolayers (SAMs) that serve as substrates for nucleation provide a way of controlling various aspects of the crystallization process with a previously unreachable precision. We focus on crystallization of calcite ($CaCO_3$) on SAMs of $HS(CH_2)_nX$ ($X = CO_2H$, CH_3, SO_3H, OH, $N(CH_3)_3Cl$) supported on Ag and Au. Fine-tuning of the crystallographic orientation of the forming crystals has been achieved by using different functional groups and metal substrates. By patterning SAMs with microregions having different nucleating activities and proper geometry, it is possible to confine crystallization to well defined, spatially delineated sites. This method provides means to fabricate arbitrarily patterned calcitic arrays with controlled density of nucleation, crystallographic orientation, and crystal sizes. The experimental conditions and the mechanisms discussed can be applied to the templated nucleation of a wide range of inorganic materials.

INTRODUCTION

Crystallization is a key process in the synthesis of many technologically important materials [1-5]. Our knowledge of basic crystallization mechanisms, e.g. what defines the location and density of nucleation, polymorph selectivity or face-selective nucleation and, after the specific nucleus is formed, what determines crystal sizes, shapes, and architecture of the formed crystals and their stability is, however, extremely limited. Better understanding of these processes is essential in engineering advanced materials with controlled properties.

In contrast, the formation of crystalline materials in nature is highly regulated [6-8]. Different organisms, irrespective of their systematic position, exercise a level of molecular control over the physico-chemical properties of minerals that is almost unthinkable in artificial processes. It is achieved by means of organized assemblies of specialized biological macromolecules. Recently, new synthetic strategies that mimic biomineralization have been developed [9-16]. These promising approaches utilize processes involving molecular recognition at the organic-inorganic interfaces. A number of elegant studies demonstrate the use of simplified molecular assemblies that resemble biological membranes, such as Langmuir monolayers [9,10,17-21], self-assembled monolayers (SAMs) [22-26], biological macromolecules [11-13] and surfactant aggregates [27-29], as the nucleation templates. Most of these studies, however, address only one "crystallization problem" at a time, e.g. oriented nucleation, polymorph specificity, or shapes of the growing crystals.

The present paper describes our approach to govern crystallization [30-31] based on engineering the nucleation site and on controlling mass transport to the surface at the micron scale, using micropatterned SAMs of alkanethiols supported on metal films. This method, which we apply to crystallization of calcium carbonate, makes it possible, for the first time, to achieve control over many of the crucial aspects of nucleation and crystal growth – oriented nucleation, location and density of nucleation, crystal sizes and patterns – in one experiment.

EXPERIMENTAL

Substrates. Silicon wafers were coated with 2 nm of Ti, to promote adhesion, and then with 50 nm of metal (Ag or Au) using an electron beam evaporator.

SAMs. SAMs of $HS(CH_2)_{15}CO_2H$, $HS(CH_2)_{11}OH$, $HS(CH_2)_{11}SO_3H$, $HS(CH_2)_{11}PO_3H_2$, $HS(CH_2)_{11}N(CH_3)_3Cl$ and $HS(CH_2)_{15}CH_3$ were formed on metal substrates by exposing the surfaces to a 10 mM solution of the thiol in ethanol for 24 hours, followed by washing with ethanol [32]. Patterned SAMs were formed using microcontact printing (μCP) [33-34]: elastomeric stamps with different relief structures were "inked" with a 10 mM solution of $HS(CH_2)_nX$ in ethanol and brought into conformal contact with gold for 10 s; the non-contact areas were then derivatized with a 10 mM solution of $HS(CH_2)_nY$ in ethanol by immersion for 1 min.

Crystallization. The substrates were placed upside-down (to ensure that only particles grown on the SAM would be bound to the surface) in 10 mM calcium chloride solution in a closed desiccator containing vials of ammonium carbonate [12, 30-31, 35-36]. All experiments were carried out at room temperature for 1 h. Precipitation of calcium carbonate results from the diffusion of carbon dioxide vapor into the $CaCl_2$ solution, according to the following reactions:

$$(NH_4)_2CO_3 \text{ (s)} \rightarrow 2NH_3 \text{ (g)} + CO_2 \text{ (g)} + H_2O;$$
$$CO_2 + Ca^{2+} + H_2O \rightarrow CaCO_3 \text{ (s)} + 2H^+;$$
$$2NH_3 + 2H^+ \rightarrow 2NH_4^+.$$

Analysis. The crystals, once formed, were examined using optical microscopy to determine the densities of nucleation and crystal sizes. The specific crystallographic orientations of the crystals relative to the interface between the SAM and the solution were analyzed using XRD in the θ–2θ scan mode. A detailed morphological analysis was undertaken to confirm the assignment of each specific crystallographic orientation and to estimate the deviation in angle of the crystals from these directions of growth (see text for details).

RESULTS AND DISCUSSION

Oriented nucleation. We have chosen the crystallization of calcite on SAMs of alkanethiols $HS(CH_2)_nX$ bearing negatively charged headgroups ($X = CO_2^-$, SO_3^-, OH) as a model system for three reasons: i) calcite has a simple structure, its crystallization is relatively easy to perform and there is an extensive background information describing this process [6, 12, 13, 19, 20, 26, 35, 37-39], ; ii) formation of crystallographically oriented, exquisitely shaped calcite crystals with unique materials properties is common in biological environments [7, 8], and is believed to be controlled by acidic macromolecules, conceivably by virtue of a match between the structures of the organic surface and that of a particular crystal plane [11, 12, 35, 36]; iii) alkanethiols self-assemble on metal substrates into highly ordered, crystalline monolayers with various structural parameters [40-44] and present, therefore, an attractive candidate for an organized organic surface to mediate crystallization [22-26, 30, 31]. Alkanethiols terminated in different functional groups assemble on Au(111) in a hexagonal overlayer characterized by the interchain distance $a = 4.97$ Å, the tilt of the chain $\alpha = 28\text{-}32°$, and the twist angle around the axis of the chain $\beta = 50\text{-}55°$ [42]. On Ag(111), X-terminated alkanethiols self-assemble on in a hexagonal array with $a = 4.77$ Å, $\alpha = 8\text{-}12°$, $\beta = 42\text{-}45°$ [43]. The presence of counterions, such as Cd^{2+} and Ca^{2+}, has been shown to induce an additional ordering of terminal acidic headgroups in SAMs [31, 45-48]. We anticipated that different SAMs – those terminated in different functional groups and even those bearing the same terminal group but supported on different metals – would induce oriented nucleation of calcite in different crystallographic directions. Indeed, crystallization experiments with CO_2^-, SO_3^- and OH-terminated SAMs supported on Au and Ag produced arrays of highly oriented calcite crystals [31], nucleated from crystallographic planes specific to each surface type (Fig. 1).

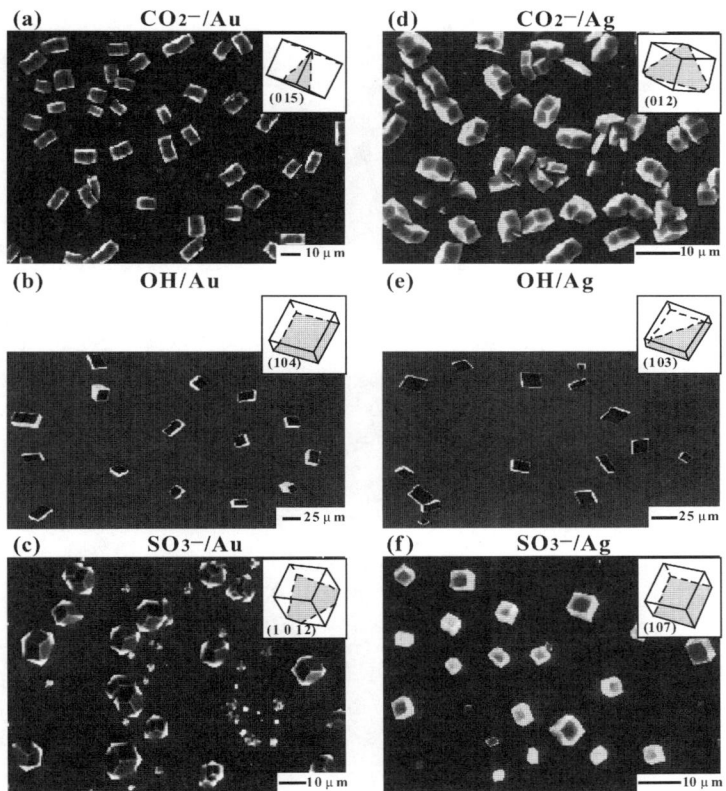

Figure 1. *Scanning electron micrographs showing the face-selective nucleation of calcite crystals mediated by SAMs. The inserts present computer generated simulations of the regular calcite rhombohedra viewed down perpendicular to the corresponding average nucleating face (shadowed).* **(a)** *CO_2^-/Au;* **(b)** *OH/Au;* **(c)** *SO_3^-/Au;* **(d)** *CO_2^-/Ag;* **(e)** *OH/Ag;* **(f)** *SO_3^-/Ag.*

The predominant nucleating planes (NP) of the calcite crystals were first determined from the XRD profiles [31]. To confirm the assignment of each specific crystallographic orientation and to estimate the angular deviation from these directions of growth, a detailed morphological analysis was performed. For each surface type, a population of 40 crystals was used in statistics. The crystals were viewed down the surface normal in a scanning electron microscope. We measured the angles between the crystal edges meeting at the upper corner of the crystal. These angles can be unequivocally related to the orientation of the regular {104} calcite rhombohedra vis-à-vis the interface, using structural relationships in the calcite unit cell [6, 37]. The morphological results were visualized using the sphere of reflection (figure 2, top).

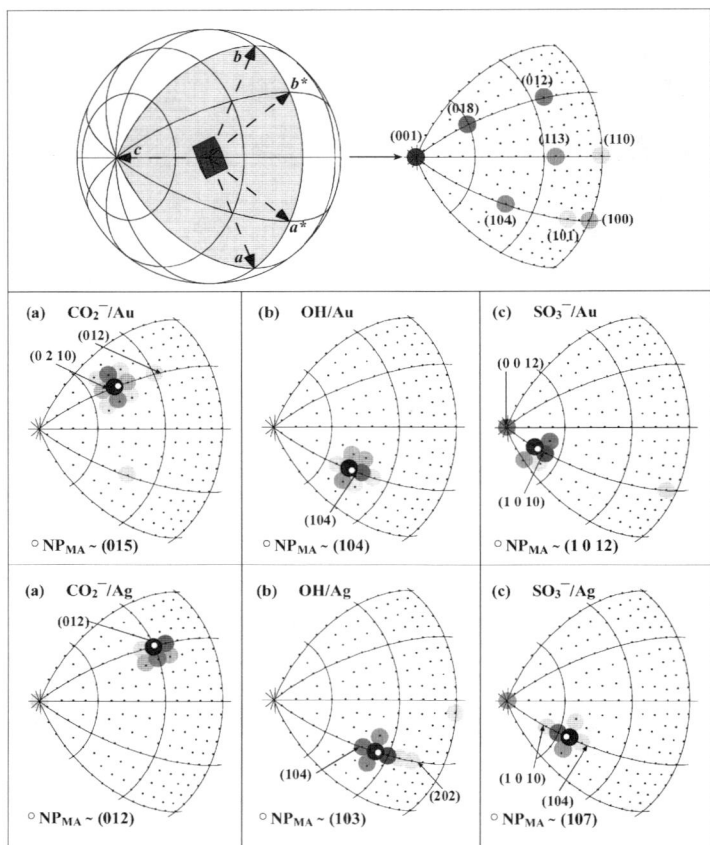

Figure 2. *Pole figure presentation of morphological data for the orientations of calcite crystals grown on different SAMs (see text for description). The indices of the crystallographic planes are indicated. (a) CO_2^- /Au; (b) OH/Au; (c) SO_3^- /Au; (d) CO_2^- /Ag; (e) OH/Ag; (f) SO_3^- /Ag.*

The nucleating planes for each surface type were plotted as circles on the sphere with a 5% gray scale; the intensity of the gray scale of the final circles is proportional to the number of crystals in the corresponding orientation [31]. The example shown in figure 2(top) corresponds to the morphological measurements of a set of 40 crystals in which one crystal nucleated from the (101) plane, two from the (110), three from the (100), four from the (113), five from the (104), six from the (012), eight from the (018) and 11 from the (001) plane. A significant clustering of the data was observed for each surface type, indicative of a high degree of orientational specificity induced by SAMs. The positions of the average nucleating planes estimated from the morphological analysis (NP_{MA}) were compared to the predominant orientations observed in the corresponding X-ray diffraction profiles (NP_{XRD}) (indicated by arrows).

Table 1 summarizes the results of oriented crystal growth on CO_2^-, SO_3^- and OH-terminated SAMs supported on Au and Ag.

Table 1. *Average nucleation plane and percentage of oriented calcite crystals grown on SAMs supported on gold and silver*

SAM	Au					Ag				
	Nucleation plane, $NP_{MA}{}^a$	Dihedral angle, δ^b	Angular deviation^c	Percentage of oriented crystals		Nucleation plane, $NP_{MA}{}^a$	Dihedral angle, δ^b	Angular deviation^c	Percentage of oriented crystals	
				From morphological analysis	From XRD analysis				From morphological analysis	From XRD analysis
CO_2^-	(015)	41°	4.3°	97	73	(012)	60°	3.6°	97	82
OH	(104)	43°	3.6°	100	91	(103)	56°	5.1°	97	81
SO_3^-	(1 0 12)	21°	3.4°	78	56	(107)	35°	2.7°	78	63

^{a} Average nucleation plane NP_{MA} corresponds to the highest density in Figure 2 (open circles).

^{b} Average dihedral angle between the NP_{MA} and (001) plane.

^{c} Standard angular deviation from the NP_{MA}.

We believe that the orientational uniformity of crystals formed on each surface is controlled by the specific interfacial structure of the oriented, homogeneous SAM. It is noteworthy that for each pair X/Au and X/Ag, the difference in crystallographic orientations of calcite crystals (~15-20°) corresponds to the difference in the tilt of the alkanethiol molecules (α) in the SAMs on Au and Ag. This relationship is illustrated in figure 3, which presents the example of oriented nucleation of calcite from the (015) and (012) crystallographic planes induced by SAMs of $HS(CH_2)_{15}CO_2H/Au$ and $HS(CH_2)_{15}CO_2H/Ag$. It is also important to note that in most cases, we *did not* observe a satisfactory match between the lattices of the SAMs and crystal planes they nucleate. On the other hand, there *always* existed a certain orientation of the functional groups in the SAM that precisely matched the orientation of the carbonate ions in the nucleated crystal (figure 3). This result provides a basis for more detailed mechanistic studies of oriented nucleation that will be addressed in our future research.

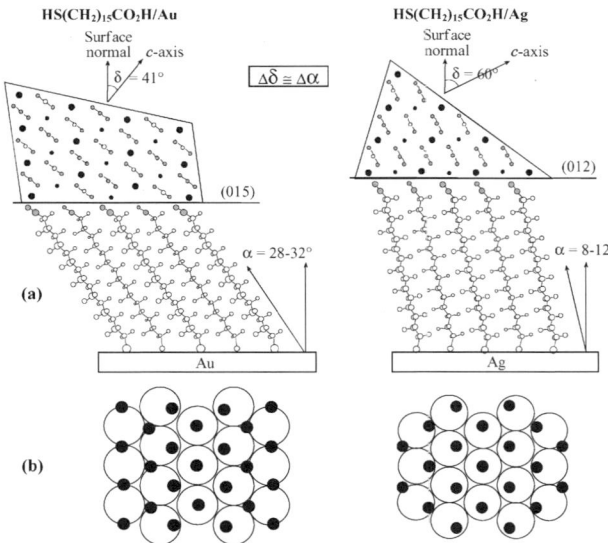

Figure 3. *Schematic presentation (to scale) of the geometrical relationship between the structure of the SAMs of HS(CH$_2$)$_{15}$CO$_2$H supported on gold (left) and silver (right) and the oriented calcite crystal they nucleate.* **(a)** *Relative orientations of the SAM and the nucleated crystals. Note the possible co-alignment of the CO$_2^-$ groups in the SAM and of the carbonates in the nucleated face;* **(b)** *Mismatch between the lattices of the CO$_2^-$ groups in the undistorted SAM (large, open circles) and the Ca^{2+} ions in the nucleated face (solid circles).*

Patterned crystal growth. Another important advantage of using SAMs of alkanethiols as nucleation templates compared to other supramolecular assemblies is that they can be easily patterned on a micron scale [33-34]. We used this property to control area-selective crystallization.

The analysis of the nucleation activity of various SAMs showed the induction time of calcite nucleation to increase and the density of nucleation to decrease in the following sequence:

$$SO_3^- - CO_2^- - OH - CH_3 - N(CH_3)_3^+.$$

Patterned SAMs of ω-terminated alkanethiols with a controlled distribution of more active nucleation sites within a less active background were formed by μCP (figure 4a). Crystallization of calcite on these substrates resulted in the formation of large-area, high-resolution inorganic replicas of the underlying organic patterns [30]. The observed patterned crystallization can be explained in terms of diffusion-limited, island-specific nucleation [30, 49]. Figure 4b describes the mechanism of the process. After the nucleation begins at rapidly nucleating SAMs, the ion flux into these regions depletes calcium carbonate concentration over slowly nucleating zones. In the region l_d where the effective concentration of the solution is below saturation (c_{sat}), nucleation does not occur. Nucleation on slowly nucleating regions is allowed only for distances from the rapidly nucleating region $x > l_d$, where $c > c_{sat}$. This mechanism was confirmed in the experiment, in which calcite crystals were grown on a methyl-terminated surface with one isolated carboxylic acid terminated region: the halo pattern clearly seen in the SEM (figure 4b) corresponds to the depletion region l_d. Therefore, if we keep the distance between rapidly nucleating regions below $2l_d$, crystallization will be entirely restricted to the rapidly nucleating regions (figure 4b).

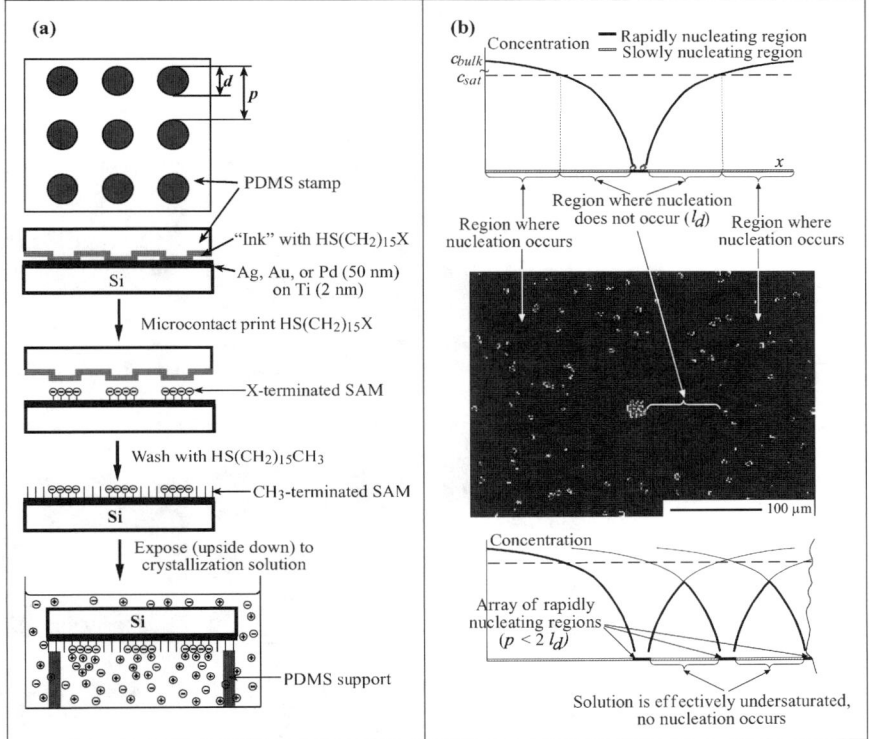

Figure 4. *(a) Schematic presentation of the experimental design of crystallization on patterned SAMs; (b) Mechanism of diffusion-limited, localized nucleation [30, 49] (see text for details).*

By adjusting various parameters of our experimental setup (concentration of the crystallizing solution, density and sizes of features in the stamp and functionality of the surface of the SAM), we can exert further control over the crystallization process: in addition to precise localization of nucleation, we can define the density of the active nucleating regions on the surface (N), the number of crystals that nucleate within each region (n), as well as their crystallographic orientation. Figure 5 presents a set of patterns of calcite crystals that exemplify the general algorithm that we used to control crystallization.

1) We choose a specific X-terminated alkanethiol/metal combination to induce the desired oriented crystallization (figure 5a).

2) For a given N, we determine the corresponding distance between the features in the stamp: $p = N^{-1/2}$.

3) By varying the concentration of the crystallizing solution, we define the range of concentrations for which $l_d > p/2$, so that inhibition of nucleation occurs over the entire slowly nucleating regions due to the mass transport to the regions of crystal growth. For example, for patterned crystallization on a SAM supported on Ag and consisted of circles of $HS(CH_2)_{15}CO_2H$

$(d = 35 \ \mu m, \ p = 100 \ \mu m)$ in the field of $HS(CH_2)_{15}CH_3$, the required concentrations for the $CaCl_2$ solution are below 100 mM (for higher concentrations, sporadic formation of crystals on the CH_3-terminated regions remote from the CO_2H-terminated sites occurred) (figure 5b).

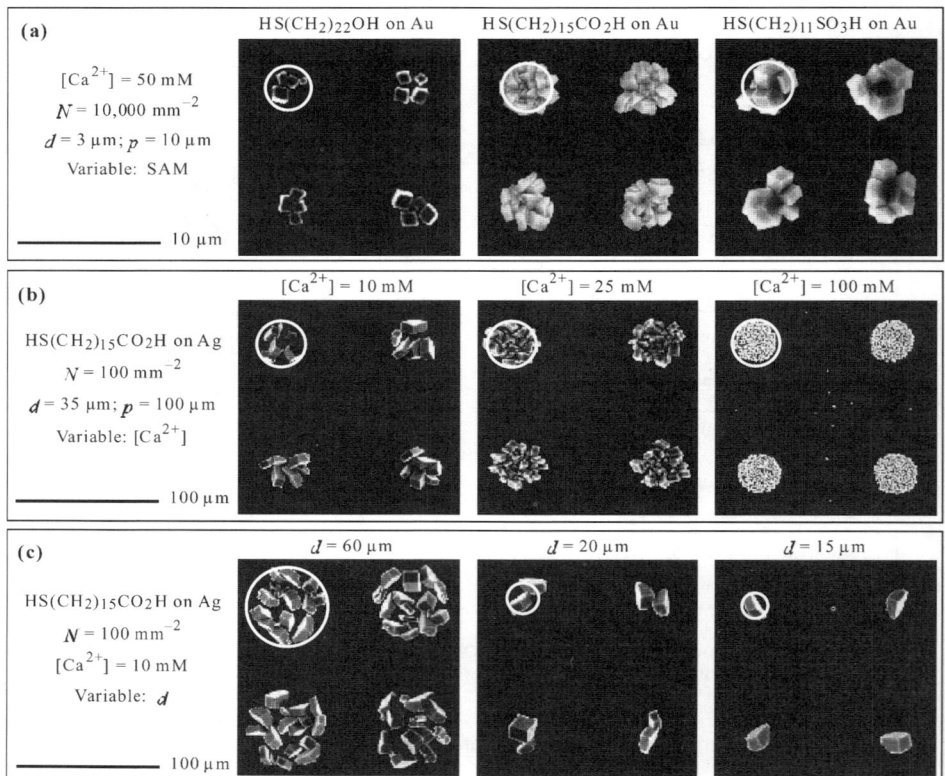

Figure 5. *Control over nucleation of calcite by micropatterned SAMs consisting of a square array of rapidly nucleating circles (SAMs of $HS(CH_2)_nX$) in a slowly nucleating background (SAMs of $HS(CH_2)_{15}CH_3$). The overlaid circles outline the geometry of the pattern. The experimental parameters – density (N) and sizes (d) of features in the stamp, concentration of the crystallizing solution, functionality of the surface of the SAM and a supporting metal – are indicated. (a) Patterned oriented crystallization induced by SAMs terminated in different functional groups. (b) Nucleation density and extent of area-selective nucleation on SAMs of $HS(CH_2)_{15}CO_2H$ supported on Ag as a function of concentration of crystallizing solution for a given distribution and size of the rapidly nucleating regions ($d = 35 \ \mu m$, $p = 100 \ \mu m$). The increase in concentration results in occasional crystallization on the methyl-terminated region, in agreement with the expected decrease in the size of the depletion zone l_d. (c) Number of crystals per active nucleation region (n) as a function of its area.*

4) At constant concentration, the number of crystals, n, within each active nucleation region appears to depend linearly upon its area [30]. This relation makes it possible to determine the optimum size, d, of the raised features in the stamp for any chosen n (figure 5c).

Using this algorithm, we have been able to determine the necessary experimental conditions – concentration of the crystallizing solution, geometric characteristics and functionality of the micropatterned SAM, supporting metal – to fabricate arbitrary patterns of oriented calcite crystals (figure 6). The crystals can be grown in dense, oriented islands (figure 6a) or in highly ordered two-dimensional arrays of single calcite crystals of uniform size and orientation (figure 6b); we can reverse the crystallization pattern and fabricate interconnected, oriented films of calcite crystals with the edge resolution of <50 nm (figure 6c); patterned growth of calcite in different crystallographic directions can be achieved by using substrates patterned with SAMs that have similar nucleating activity, but induce nucleation from different crystallographic planes (figure 6d).

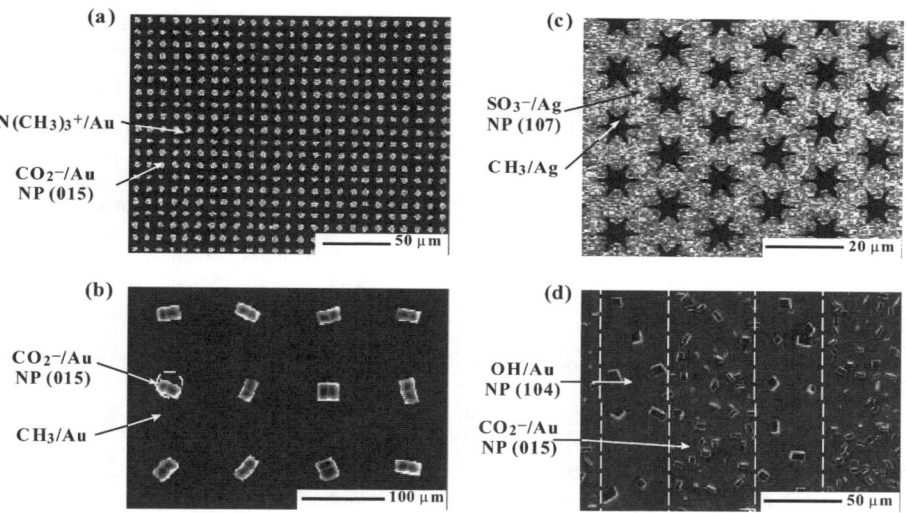

Figure 6. *Scanning electron micrographs showing the controlled crystallization of calcite on patterned SAMs. (a) A square array of densely nucleated calcite crystals; (b) A square array of discrete, oriented calcite crystals; (c) Complex pattern of an interconnected, oriented calcitic film; (d) Simultaneous patterned growth of calcite in two crystallographic directions.*

CONCLUSIONS

Diffusion-controlled growth of calcite in nature is a well-known phenomenon [6-8]. Formation of intricate biogenic calcitic structures usually occurs from organized macromolecules in specific microenvironments that provide the directional flux of ions towards the growing crystals [7, 8]. The power of the presented bio-inspired approach to artificial crystallization is, therefore, based on our ability to govern mass transport to different regions of the surface at a micron scale and to control the molecular structure and microenvironment of the nucleation site, by patterning ω-terminated SAMs into regions with different nucleating activity. Using crystallization of calcite as an example, we demonstrate that we can regulate the precise location, density and area of the regions active for nucleation, and thereby the number, size and crystallographic orientation of the crystals that nucleate in each part of the surface.

We conclude, therefore, that the combination of three major ideas – (i) the ability of SAMs to nucleate growth from a single crystallographic plane, (ii) patterning SAMs with microregions having different nucleating activity, and (iii) taking advantage of mass transport and the proper size of the pattern, so that the ion flux into the regions of crystal growth keeps the concentration of the crystallizing solution over slowly nucleating regions low enough that nucleation effectively never happens, – provides a simple and convenient route to control nucleation. We believe that this method could find applications in engineering of oriented inorganic materials with complex form, such as ceramics or semiconductors whose mechanical, optical and electrical performance can be regulated by controlling the sizes, distribution, and morphology of the constituent crystals.

ACKNOWLEDGMENTS

I thank Dr. A. J. Black and Prof. G. M. Whitesides for their valuable contribution to this project.

REFERENCES

1. A. H. Heuer, D. J. Fink, V. J. Laraia, J. L. Arias, P. D. Calvert, K. Kendall, G. L. Messing, J. ̄ ̄ ̄ ̄ D. H. Thompson, A. P. Wheeler, A. Veis and A. I. Caplan, *Science*, **255**, 1098 (1992).
2. C. B. Murray, C. R. Kagan and M. G. Bawendi, *ibid.*, **270**, 1335 (1995).
3. S. I. Stupp and P. V. Braun, *ibid.*, **277**, 1242 (1997).
4. B. J. J. Zelinsky, C. J. Brinker, D. E. Clark and D. R. Ulrich, Eds., *Better Ceramics Through Chemistry* (Materials Research Society. Pittsburgh, 1990).
5. S. Mann and G. A. Ozin, *Nature*, **382**, 313 (1996).
6. F. Lippmann, *Sedimentary Carbonate Minerals* (Springer-Verlag, Berlin, 1973).
7. H. A. Lowenstam and S. Weiner, *On Biomineralization* (Oxford Univ. Press, 1989).
8. S. Mann, J. Webb and R. J. P. Williams, Eds., *Biomineralization. Chemical and Biological Perspectives* (VCH, Weinheim, 1989).
9. E. M. Landau, M. Levanon, L. Leiserowitz, M. Lahav and J. Sagiv, *Nature*, **318**, 353 (1985).
10. E. M. Landau, S. G. Wolf, M. Levanon. L. Leiserowitz, M. Lahav and J. Sagiv, *J. Am. Chem. Soc.*, **111**, 1436 (1989).
11. L. Addadi and S. Weiner, *Proc. Natl. Acad. Sci. USA*, **82**, 4110 (1985).
12. L. Addadi, J. Moradian, E. Shay, N. G. Maroudas and S. Weiner, *ibid.*, **84**, 2732 (1987).
13. A. M. Belcher, R. J. Christensen, P. K. Hansma, G. D. Stucky and D. E. Morse, *Nature*, **381**, 56 (1996).
14. M. Alper, P. D. Calvert, R. Frankel, P. C. Rieke and D. A. Tirrell, Eds., *Materials Synthesis Based on Biological Processes* (Materials Research Society, Pittsburgh, 1991).
15. S. Mann, *Nature*, **365**, 499-505 (1993).

16. S. Mann, D. D. Archibald, J. M. Didymus, T. Douglas, B. R. Heywood, F. C. Meldrum and N. J. Reeves, *Science*, **261**, 1286 (1993).
17. X. K. Zhao and J. H. Fendler, *J. Phys. Chem.*, **95**, 3716-3723 (1991).
18. B. R. Heywood and S. Mann, *J. Am. Chem. Soc.*, **114**, 4681 (1992).
19. S. Mann, B. R. Heywood, S. Rajam and J. D. Birchall, *Nature*, **334**, 692 (1988).
20. P. W. Carter and M. D. Ward, *J. Am. Chem. Soc.*, **115**, 11521 (1993).
21. L. M. Frostman and M. D. Ward, *Langmuir*, **13**, 330 (1997).
22. B. C. Bunker, P. C. Rieke, B. J. Tarasevich, A. A. Campbell, G. E. Fryxell, G. L. Graff, L. Song, J. Liu, J. W. Virden and G. L. McVay, *Science*, **264**, 48 (1994).
23. S. Feng and T. Bein, *Nature*, **368**, 834 (1994).
24. L. M. Frostman, M. M. Bader and M. D. Ward, *Langmuir*, **10**, 576 (1994).
25. V. K. Gupta and N. L. Abbott, *Science*, **276**, 1533 (1997).
26. A. Berman, D. J. Ahn, A. Lio, M. Salmeron, A. Reichert and D. Charych, *Science*, **269**, 515 (1995).
27. D. D. Archibald and S. Mann, *Nature*, **364**, 430 (1993).
28. D. Walsh, J. D. Hopwood and S. Mann, *Science*, **264**, 1576 (1994).
29. T. Douglas, D. P. E. Dickson, S. Betteridge, J. Charnock, C. D. Garner and S. Mann, *ibid*, **269**, 54 (1995).
30. J. Aizenberg, A. J. Black and G. M. Whitesides, *Nature*, **398**, 495 (1999).
31. J. Aizenberg, A. J. Black and G. M. Whitesides, *J. Am. Chem. Soc.*, **121**, 4500 (1999).
32. N. B. Larsen, H. Biebuyck, E. Delamarche and B. Michel, *J. Am. Chem. Soc.*, **119**, 3017 (1997).
33. A. Kumar, H. A. Biebuyck and G. M. Whitesides, *Langmuir*, **10**, 1498 (1994).
34. A. Kumar, N.A. Abbott, E. Kim, H.A. Biebuyck and G. M. Whitesides, *Acc. Chem. Res.*, **28**, 219 (1995).
35. A. Berman, L. Addadi and S. Weiner, *Nature*, **331**, 546 (1988).
36. S. Albeck, J. Aizenberg, L. Addadi and S. Weiner, *J. Am. Chem. Soc.*, **115**, 11691 (1993).
37. D. D. Archibald, S. B. Qadri and B. P. Gaber, *Langmuir*, **12**, 538 (1996).
38. G. Falini, S. Albeck, S. Weiner and L. Addadi, *Science*, **1996**, *271*, 67-69.
39. H. H. Teng, P. M. Dove, C. A. Orme and J. J. De Yoreo, *Science*, **282**, 724 (1998).
40. A. Ulman, *An Introduction to Ultrathin Organic Films: From Langmuir-Blodgett to Self-Assembly* (Academic Press, San Diego, 1991).
41. P. E. Laibinis and G. M. Whitesides, *J. Am. Chem. Soc.*, **114**, 1990 (1992).
42. R. G. Nuzzo, L. H. Dubois and D. L. Allara, *J. Am. Chem. Soc.*, **112**, 558 (1990).
43. P. E. Laibinis, G. M. Whitesides, D. L. Allara, Y. T. Tao, A. N. Parikh and R. G. Nuzzo, *J. Am. Chem. Soc.*, **113**, 7152 (1991).
44. N. Camillone, C. E. D. Chidsey, G. Liu and G. Scoles, *J. Chem. Phys.*, **98**, 4234 (1993).
45. F. Leveiller, D. Jacquemain, M. Lahav, L. Leiserowitz, M. Deutsch, K. Kjaer and J. Alsnielsen, *Science*, **252**, 1532 (1991).
46. J. A. Zasadzinski, R. Viswanathan, L. Madsen, J. Garnaes and D. K. Schwartz, *Science*, **263**, 1726 (1994).
47. C. Böhm, F. Leveiller, D. Jacquemain, H. Mohwald, K. Kjaer, J. Alsnielsen, I. Weissbuch and L. Leiserowitz, *Langmuir*, **10**, 830 (1994).
48. J. Li, K. S. Liang, G. Scoles and A. Ulman, *Langmuir*, **11**, 4418 (1995).
49. A.-L. Barabási and H. E. Stanley, *Fractal Concepts in Surface Growth* (Cambridge Univ. Press, 1995).

Mat. Res. Soc. Symp. Vol. 620 © 2000 Materials Research Society

Angular Fourier Mapping;
Highlighting lattice structures without destroying original data

Johannes H. Kindt*, James B. Thompson*, George T. Paloczi*, #, Martina Michenfelder§,
Bettye L. Smith*, §, Galen Stucky¶, Daniel E. Morse§, and Paul K. Hansma*
*Department of Physics, University of California Santa Barbara, Santa Barbara, CA 93106, USA.
#Now at Department of Physics, California Institute of Technology, Pasadena, CA 91125, USA.
§Department of Molecular, Cellular and Developmental Biology, University of California Santa Barbara, Santa Barbara, CA 93106, USA.
¶Department of Chemistry, University of California Santa Barbara, Santa Barbara, CA 93106, USA.

ABSTRACT

A two-dimensional Fourier transformation, FT, is used to isolate two different lattice structures within one scanning probe microscope, SPM, image. The isolated structures are then used to create a two-color map that encodes the presence of these structures within the image. The color map is normalized in brightness and then used to color-code the original black and white SPM data. The distribution of different structures becomes obvious, while all original brightness information is preserved in this combined image.

INTRODUCTION

Lattice-scale SPM images are often obscured by noise. Digital filtering can be employed to remove that noise and make features such as step-edges more clearly visible. The most broadly used filtering technique involves a conversion of the image into the frequency domain (FT), selection of image frequencies and exclusion of noise frequencies, and an inverse transformation [1]. While this is a powerful method, many researchers avoid using it because its effects on the original data are hard to predict. In principle, a carefully tuned Fourier filter can turn white noise into any image desired.

The algorithm proposed here employs Fourier filtering but it puts restrictions on the filtering parameters, making the result more predictable. Furthermore, the original black/white coded information is not changed, but only color-coded with the result of the digital filter. To return to the original image the color coding can be removed by anyone, e.g. using a Xerox copier with a flat spectral response. The main application for this method is to visually separate lattice structures of different periodicity and orientation.

Algorithm

The Angular Fourier Mapping algorithm can be divided into the following six consecutive steps:
1. Two-dimensional FT of the original SPM image

Original Image:	$O(x,y)$	Fig. 1
2D FT:	$F(x,y) = (O(x,y))$	Fig. 2

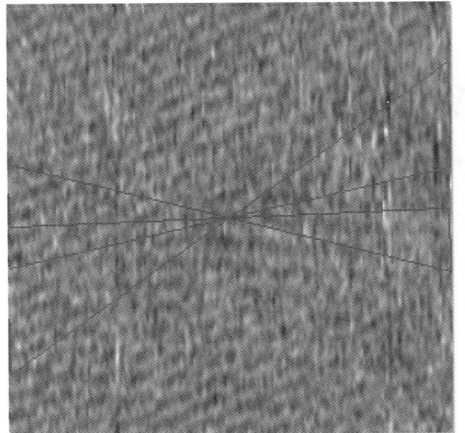 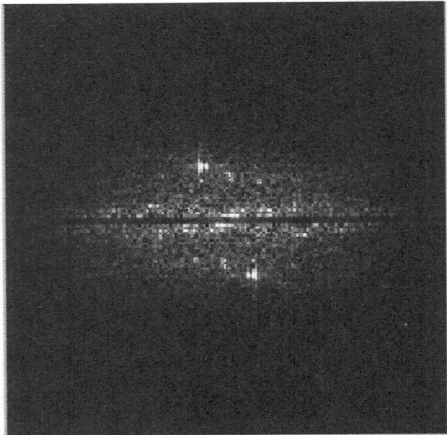

Figure 1. *Original height image of a transition in lattice structure, observed by atomic force microscopy in liquid*   **Figure 2.** *Two-dimensional FT of the original data.*

2. User selection of two lattices by periodicity and orientation

The FT translates lattice periodicity into radii from the center. Angles translate into angles. A lattice is fully defined by its period and orientation. A lattice in the original image appears as a bright maximum in the Fourier image.

Two different lattices are now selected, each by choosing a region in Fourier space bounded by a set of radii and angles that enclose the maximum.

$$\rightarrow \qquad A(r_1, r_2, \varphi_1, \varphi_2) \qquad and \qquad B(r_3, r_4, \varphi_3, \varphi_4)$$

Since angles translate into angles, they can simultaneously be displayed and verified in the original image. Then, everything outside the so selected regions of the Fourier image is masked:

$$F_A(x',y') = \begin{cases} F(x',y'), & x',y' \text{ inside } A \\ 0, & otherwise \end{cases}$$

$$F_B(x',y') = \begin{cases} F(x',y'), & x',y' \text{ inside } B \\ 0, & otherwise \end{cases}$$

3. Inverse FT of the masked areas:

The selected areas are converted back into space coordinates. They now contain only contain the main components of the selected lattice structures.

$$O_A(x,y) = \quad {}^{-1}(F_A(x',y')), \qquad Fig.\ 3a$$
and
$$O_B(x,y) = \quad {}^{-1}(F_B(x',y')) \qquad Fig.\ 3b$$

Figure 3a/b. *Inverse FT of the selected regions in the Fourier image.*

4. Turn amplitude into color intensity:

Now, the local Amplitude of these filtered lattice structure components is analyzed. Since amplitude is defined by the peaks of a function, it is not an entirely local phenomenon. A good way to define amplitude for every location in the picture is to average the absolute values of all neighboring pixels.

$$\vec{m}_A(x,y) = average(|O_A|, x, y, r_{blur}) \cdot \vec{color}_A , \qquad \text{Fig. 4a}$$

$$\vec{m}_B(x,y) = average(|O_B|, x, y, r_{blur}) \cdot \vec{color}_B \qquad \text{Fig. 4b}$$

average is a $1/r^2$ - weighted pixel-by-pixel average with the maximum distance $r_{average}$ from the original pixel:

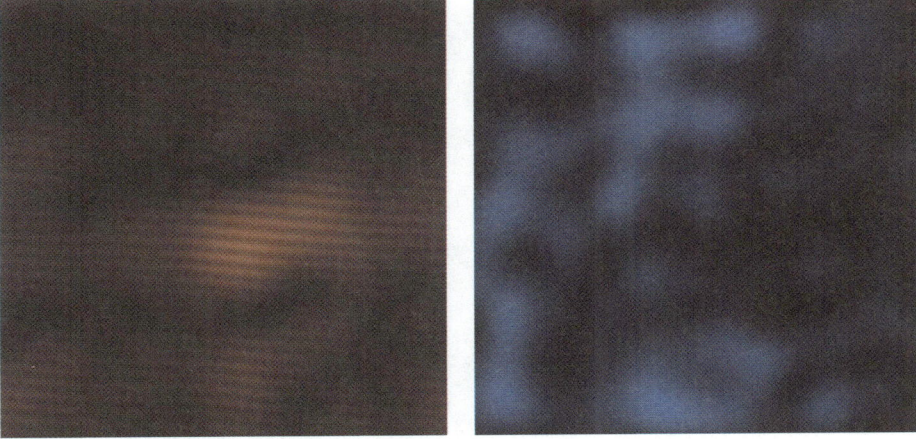

Figure 4a/b. *Amplitude to Brightness of color conversion of Figure 3a/b.*

$$average(O_A,x,y,r_{average}) = \sum_{i=x-r_{average}}^{x+r_{average}} \sum_{j=y-r_{average}}^{y+r_{average}} \frac{O_A(x,y)}{\sqrt{i^2+j^2}} \cdot \left(\sum_{i=x-r_{average}}^{x+r_{average}} \sum_{j=y-r_{average}}^{y+r_{average}} \frac{1}{\sqrt{i^2+j^2}} \right)^{-1}$$

$r_{average}$ is set to be the same as the lattice spacing.

5. Compose the color maps and normalize brightness
   In this step, the two color maps are merged together by averaging them pixel by pixel.

$$\overrightarrow{mixedmap}(X,Y) = \frac{\overrightarrow{m_A}(x,y) + \overrightarrow{m_B}(x,y)}{2}$$

Pixel Colors are written as vectors. In the RGB system, the color vector has the components:

$$\overrightarrow{Color} = (R_{ed} \quad G_{reen} \quad B_{lue})$$

The origin of this system represents black. The three unity vectors represent red, green and blue. Adding the three unity vectors results in white (additive color mixing). The distance from the origin is the brightness: $\quad Brightness = \left|\overrightarrow{Color}\right|$

In order to not affect the brightness of any pixel in the image later, all pixels of the color map are normalized to one brightness:

$$\overrightarrow{normmap}(x,y) = \frac{\overrightarrow{mixedmap}(x,y)}{\left|\overrightarrow{mixedmap}(x,y)\right|} \qquad\qquad \text{Fig. 5}$$

The map now contains only color information, its brightness is uniformly one.

6. Superimpose the composed color map and the original SPM image
   Finally, the brightness of every pixel in the original image is multiplied with its corresponding pixel in the color map.

$$\overrightarrow{mapimage}(x,y) = \overrightarrow{normmap}(x,y) \cdot O(x,y) \qquad\qquad \text{Fig. 6}$$

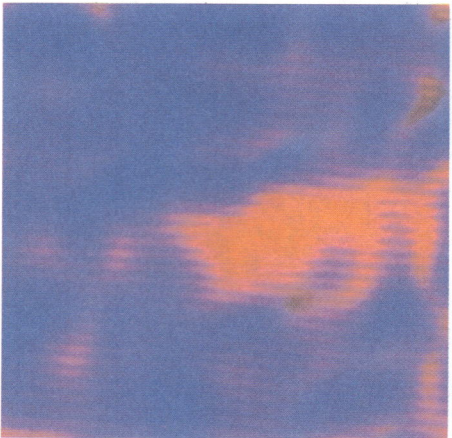 

Figure 5. Brightness-normalized superposition of Figure 4a and 4b.

Figure 6. Original image, color-coded with color map from Figure 5

The result is an image that contains brightness information only from the original image, and color coding from the Fourier filter.

Experimental details

The Angular Fourier Mapping technique described was implemented in Delphi 4 for Windows, and executed on a Pentium II-200MHz. It can be obtained from [2].
The data set used to demonstrate this technique has been acquired using a Digital Instruments Multimode Atomic Force Microscope, AFM, in liquid. Previous studies have shown that the mixture of soluble proteins found in abalone nacre cause the nucleation and growth of aragonite needles on the $(10\underline{1}4)$ faces of calcite seed crystals in supersaturated calcium carbonate solutions [3]. Similar control of calcium carbonate polymorph has also been reported for soluble proteins extracted from other mollusc species [4-6]. Previously, the AFM has been used to study crystal growth in-situ [7-10], and to observe the atomic lattice of calcite [11-14]. In particular, proteins isolated from abalone nacre were shown to affect the shape of calcite growth steps, roughen the calcite surface [15] and change the observed lattice structure on the calcite surface [16]. Here we use data from the last study to demonstrate the Angular Fourier Mapping technique described. The FT and inverse FT were implemented using the Fast Fourier algorithm (FFT) [17].

Discussion

An AFM image (Fig.1) of two interlocking lattice structures is analyzed using Angular Fourier Mapping. In Fourier space (Fig.2), the maxima that relate to the different lattice structures are prominent. Now, one picks two regions in Fourier space. If each of these regions contains one of the maxima, the following inverse FT will produce images that contain the two related lattice structures(Fig.3a/b). Please note that selecting the two regions in Fourier space, by setting an upper and a lower limit for both orientation and periodicity, is the only degree of freedom, and the only influence one has on the final result of the algorithm. This also means that the algorithm does not find lattice structures by itself, it only presents a tool to better identify and then highlight them. The so filtered structures are then used to modulate the brightness of an assigned color (Fig.4a/b). The resulting color maps are merged together, and the brightness of every pixel is normalized(Fig.5). If a pixel contains color from both color maps, the result will be a mixed color. If a pixel did not contain any color information, normalizing will turn it gray. Finally, the original grayscale data is color coded with the brightness normalized color map (Fig.6). This adds color information about structural components that fall into ones specifications regarding orientation and periodicity. The brightness information from the original grayscale image is preserved.

Conclusions

The algorithm presented here has proven very useful and intuitive in our recent experiments that were also the source of the data set used here as an example. Furthermore, it shows how to avoid the risks toward the originality of data by numerical data processing: The processing parameters are limited to a small, intuitively accessible set, and information is only added, not changed, and can easily be removed to restore the original image.

References

[1] Park, S.-I. and C. F. Quate. 1987. Digital filtering of scanning tunneling microscope images. *J. Appl. Vac. Phys.* 312-14

[2] Johannes H. Kindt. Lattice Paint. *http://hansmalab.physics.ucsb.edu/latticepaint/*
At the date this paper was written, URLs or hotlinks referenced herein were deemed to be useful supplementary material to this paper. Neither the author nor the Materials Research Society warrants or assumes liability for the content or availability of URLs referenced in this paper.

[3] Belcher, A. M., X. H. Wu, R. J. Christensen, P. K. Hansma, G. D. Stucky, and D. E. Morse. 1996. Control of crystal phase switching and orientation by soluble mollusc-shell proteins. *Nature.* 381:56-58.

[4] Falini, G., S. Albeck, S. Weiner, and L. Addadi. 1996. Control of aragonite polymorphism by mollusk shell macromolecules. *Science.* 271:67-69.

[5] Samata, T., N. Hayashi, M. Kono, K. Hasegawa, C. Horita, and S. Akera. 1999. A new matrix protein family related to the nacreous layer formation of Pinctada fucata. *Febs Lett.* 462:225-229.

[6] Kono, M., N. Hayashi, and T. Samata. 2000. Molecular mechanism of the nacreous layer formation in Pinctada maxima. *Biochem. Bioph. Res. Co.* 269:213-218.

[7] Kuznetsov, Y. G., A. J. Malkin, W. Glantz, and A. McPherson. 1995. In situ atomic force microscopy studies of protein and virus crystal growth mechanisms. *In* Sixth International Conference, Hiroshima, Japan. 63-73.

[8] Paloczi, G. T., B. L. Smith, P. K. Hansma, D. A. Walters, and M. A. Wendman. 1998. Rapid imaging of calcite crystal growth using atomic force microscopy with small cantilevers. *Appl. Phys. Lett.* 73:1658-1660.

[9] Teng, H. H., P. M. Dove, C. A. Orme, and J. J. De Yoreo. 1998. Thermodynamics of calcite growth: baseline for understanding biomineral formation. *Science.* 282:724-727.

[10] Land, T. A., T. L. Martin, S. Potapenko, G. T. Palmore, and J. J. De Yoreo. 1999. Recovery of surfaces from impurity poisoning during crystal growth. *Nature.* 399:442-445.

[11] Hillner, P. E., A. J. Gratz, S. Manne, and P. K. Hansma. 1992. Atomic-scale imaging of calcite growth and dissolution in real time. *Geology.* 20:359-362.

[12] Ohnesorge, F., and G. Binnig. 1993. True atomic-resolution by atomic force microscopy through repulsive and attractive forces. *Science.* 260:1451-1456.

[13] Stipp, S. L. S., C. M. Eggleston, and B. S. Nielsen. 1994. Calcite surface structure observed at microtopographic and molecular scales with atomic force microscopy (AFM). *Geochim. Cosmochim. Ac.* 58:3023-3033.

[14] Liang, Y., A. S. Lea, D. R. Baer, and M. H. Engelhard. 1996. Structure of the cleaved $CaCO_3$ ($10\bar{1}4$) surface in an aqueous environment. *Surf. Sci.* 351:172-82.

[15] Walters, D. A., B. L. Smith, A. M. Belcher, G. T. Paloczi, G. D. Stucky, D. E. Morse, and P. K. Hansma. 1997. Modification of calcite crystal growth by abalone shell proteins: an atomic force microscope study. *Biophys. J.* 72:1425-33.

[16] Thompson, J. B., G. T. Paloczi, J. H. Kindt, M. Michenfelder, B. L. Smith, G. Stucky, D. E. Morse, and P. K. Hansma. 2000. Direct observation of the transition from calcite to aragonite growth as induced by abalone shell proteins. *Biophys. J.* In Press.

[17] Cooley, J. W. And J.W. Tukey. 1965. An algorithm for the machine calculation of complex Fourier series. *Math. Comp.* 19:297-301.

The authors gratefully acknowledge the support of the Army Research Office, MURI Program (Grant No. DAAH04-96-1-0443), the Office of Naval Research (Grant No. N00014-93-10584), the Materials Research Laboratory Program of the National Science Foundation (Grant No. DMR-9632716), the National Science Foundation (Grant No. DMR-9622169). The U.S. Government is authorized to reproduce and distribute copies for governmental purposes.

Mat. Res. Soc. Symp. Vol. 620 © 2000 Materials Research Society

Biogenic Inspiration For The Controlled Nucleation And Growth Of Inorganic Materials

Brigid R Heywood, Susan Hill, Kate Pitt, Paul Tibble, and Stuart Williams
Crystal Science Group, Lennard-Jones Laboratories,
School of Chemistry & Physics,
Keele University, Keele,
Staffs. ST5 5BG, UK

ABSTRACT

The development of effective protocols for the control of crystal structure, size and morphology attracts considerable interest given the requirement for particles of modal size and shape in many areas of particle processing and the importance of crystallochemical selectivity in determining the exploitable properties of crystalline solids. In biological systems there are many examples of advanced "crystal engineering" in which materials are deposited in a highly controlled manner to produce crystal phases that are unique with respect to their structure, habit, uniformity of size and texture. A review of biomineralisation will show that while a complex array of strategies have evolved for regulating crystal growth, one feature is common to the biological paradigm. Interactions between supramolecular organic structures and the nascent inorganic solids play a fundamental role in controlling the deposition of the biominerals and ordering the assembly of these units into hierarchical structures. In order to gain a better understanding of the molecular recognition events, which take place at the organic-inorganic interface, a bio-inspired crystal chemical approach has been adopted. For this work organised organic assemblies (e.g. surfactant aggregates, peptide mimics, dendrimers) of precise molecular design (*head group identity, packing conformation, primary sequence etc.*) are being assayed for their effectiveness in controlling the nucleation and growth of crystals. It is evident from these studies that the chemical organisation of the polymeric microenvironment operates at the molecular level to control certain aspects of the nucleation, growth and stabilisation of inorganic particles. By systematically changing the molecular motif of the organic template we have established that the size, crystallographic orientation, growth and assembly of the mineral phase can be tailored to function. These results have relevance not only to our understanding of biomineralisation but also suggest a multiplicity of exploitable opportunities for the engineering of crystals.

INTRODUCTION

Of the many roles ascribed to metals in biological systems perhaps non is more challenging to the imagination than the formation of solid phase inorganic deposits. Biological minerals formed from calcium phosphates, carbonates, oxalates and sulphates are widespread amongst plants and animals, as are deposits of iron oxides and sulphides, barium and strontium sulphates and silica [1]. Even precipitates of non-essential elements (silver, gold and uranium) have been identified in the cell walls of a some of bacteria [2,3] In addition to the more obvious examples of biominerals in structural/defensive applications (*bones, teeth, shells etc.*), devices for inertial detection, magnetic sensing, optical transmittance and homeostatic regulation are included amongst the many varied uses for which they are utilised *in vivo*. Thus, biological minerals arrest our interest not only because of their novel and elegant forms but also because they are employed for specialist functions, which require materials of specific size, structure, shape, orientation and organisation to optimise performance. Thus, a casual appraisal of biomineralisation makes it clear that 'nature' has evolved unique crystallisation strategies that circumvent the accepted thermodynamic directives normally controlling crystal nucleation and growth. The result is the facilitated fabrication of inorganic materials tailored for precise functional use [4]. The elucidation of these strategies so that they can be mimicked, adapted and controlled to produce novel materials with enhanced properties, (and the added benefit of environmental compatibility), is now the common goal of materials scientists, gene technologists, microbiologists, protein chemists, structural engineers and crystal chemists.

To the crystal scientist, a review of biomineralisation shows that while a complex array of strategies have evolved to regulate the formation of bio-crystalline phases one feature is common to the biological paradigm. Specific and selective interactions between organised biopolymeric assemblies (e.g. collagen-based matrices, β-pleated polypeptide sheets, and proteolipid membranes), and the nascent inorganic solids are fundamental to the crystallisation process [5]. Indeed, it is now clear that the events of nucleation and growth, the control of crystal morphology and the aggregation (assembly) of the resulting mineral units into functional hierarchical micro-architectures, are all regulated by these bio-organic matrices.

Circumstantial evidence suggests that the selective interaction of the precipitating cations and anions with a unique catalogue of bio-organic molecules which cycle through during mineralisation [7] simulates the lattice geometry (*co-ordination chemistry, hydration profile etc.*) of the ions in the mineral phase and potentiates the mineralisation event. The observation that bioorganic molecules with functionalities that mimic the oxyanions of the contiguous biomineral are frequently isolated from mineralised tissues gives further support to this argument [8]. For example, carboxylate-rich acidic proteins are commonly associated with calcium carbonate biominerals [8], whereas the species isolated from calcium phosphate biominerals are heavily phosphorylated [9] and the organic matrices associated with iron oxide, deposition, silica or ice formation are distinguished by the abundance of hydroxyl groups [10].

On the basis of these observations and evidence obtained from comparative structural analyses of several biominerals, a purely epitactic relationship was proposed. In this model, which borrows much from geological crystal growth phenomena [11], a direct geometric correspondence between the lattice co-ordinates defining the principal face of the biomineral and specific conformational features of the organic matrix was proposed as the primary regulator of crystal formation. However, the failure of many *in vitro* crystallisation assays utilising macromolecules isolated from mineralising matrices to verify this model raised questions about its validity and relevance. Moreover, recent genetic and biochemical studies concerning the primary structure of matrix constituents (principal functional residues, degree and type of post-translational modifications, viz. *phosphorylation, glycosylation, sulphation etc*) prompted the evolution of an interfacial (*organic-inorganic*) model encompassing a more complex interplay of molecular parameters. These would include, for example, geometric complementarity, charge localisation, the chemistry of the hydration layers and the importance of steric motifs as the defining elements of the organic template. Moreover, these parameters would be the key agents controlling the subsequent patterning of the precipitating mineral phase through the agency of the unique molecular motifs dominating the structure of the polymeric template. The next the question was how to verify and expand this more complex model?

One approach involves the direct probing of mineralising systems *in vivo* using the sophisticated repertoire of techniques currently available to molecular biologists (*cDNAs, expression cloning, in situ hybridisation and developmental switching of gene expression*) [12].

For solid state chemists, engineers and material scientists the challenge rests with combining in depth studies of the mineral phase with the development of effective and reproducible '*biomimetic*' crystallisation strategies. The purely synthetic challenge of designing polymers with the appropriate architecture and chemistry surface functionalities that are active with respect to crystal nucleation and growth was successfully taken up by some groups [13-15]. The further development of this work from the phenomenological to a predictive science has proven more challenging since the molecular identity of the active sites is difficult to characterise.

As an alternative, the feasibility of using supramolecular assemblies of organic molecules as model templates to control nucleation and growth is under consideration. This approach to the engineering of crystals has the advantage of providing a route to control the interfacial chemistry by molecular recognition through the synthesis of tailored organic templates (*head group identity, polarity, spatial ordering, etc.,* in order to initiate and regulate crystal growth. By adopting ideas and concepts derived from biomineralisation this novel interdisciplinary approach to the engineering of crystals has brought about the successful integration of organic supramolecular chemistry and molecular self-assembly with the classical elements of inorganic synthesis and materials science to open new horizons in crystal growth and materials research. In this paper, *examples* drawn from recent work investigating the potential of simple anionic species to control and direct crystal growth are reported. The facility of more complex molecular species to dictate the nucleation and growth of inorganic phases has also been investigated; the goal throughout being to rationalise, in a systematic way, the fundamental agents controlling the nucleation and growth processes.

ONE DIMENSIONAL TEMPLATES – *CONTROL OF CRYSTAL GROWTH*

As noted earlier one of the critical observations to arise from analyses of the matrices of mineralising tissues was the observation that many of these bio-organic molecules present with functionalities (e.g. sulphate, phosphate, carboxylate, amine) which mimic the oxyanions of the contiguous biomineral. Equally, the potential importance of soluble inorganic additives (e.g. magnesium, zinc) merits consideration given the known associations of such species with cellular function and matrix formation [16]. It seemed relevant, therefore, to review the impact of simple chemical analogues of these biological species upon the formation of crystal growth.

The presence of soluble organic additives (e.g. α,ω dicarboxylates) has a significant effect upon the morphology of calcite crystals [17]. Indeed, these organic acids will transform calcite from its normal rhombohedral ($\{10.4\}$) habit into spindle-shaped crystals elongated along the c axis with curved $\{10.0\}$ prismatic faces (see Figures 1 a & b). The planar bidentate COO⁻ groups of the dicarboxylate mirror the surface incorporation of CO_3^- into the $\{10.0\}$ faces; this stereochemical relationship is compounded by a potential geometric match between adjacent layers of carbonates in the designated face of the crystal lattice and the two carboxylate residues in the additive molecules. This type of *substrate recognition* can also be invoked to explain the specific interaction of other soluble additives with calcite (see Figure 1c). Interestingly, many biogenic crystals also display related morphologies [18], and when characterised they are found to be associated intimately with organic molecules of comparable chemical identity [19].

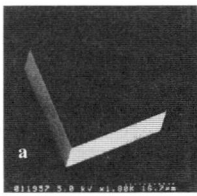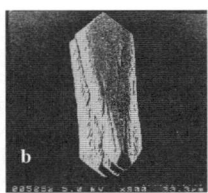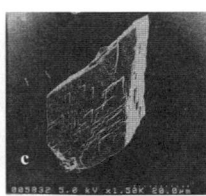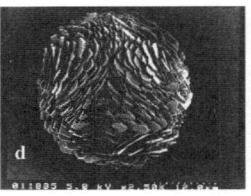

Figure 1. (a) Calcite control expressing {10.4} equilibrium morphology; (b) calcite modified by the addition of orthophosphate (Ca:additive = 500:1) - the {041} faces now dominate; (c) the addition of Mg²⁺ (Ca:Mg = 1:1) favours the {011} faces; (d) complex intergrown, contact twins dominated by {10.0} & (001) faces arise from of Zn²⁺ (Ca:Zn = 5000:1) doping of the lattice.

Equally, the selective doping of the calcite lattice with cations, which approximate the co-ordination environment of calcium also, provokes the specific and selective habit modification of the crystalline phase (see Figure 1d). Here the effects of zinc incorporation are reported given the known association of this metal with the processes of matrix fabrication [16]. Other bioinorganic cations that evidence specific activity with respect to the growth of calcite are manganese, cobalt and magnesium. By adopting a predictive approach to the design and selection of these additives it is possible to generate a directory of crystalline solids, many of which portray unique chemico-physical properties *(e.g. zeta potential, optical reflectance, acid/base dissolution profiles)* which can be exploited through a range of applications.

When small, but more complex oligomeric species are employed as solution phase additives similar effects are noted. Here we have investigated the activity of short homopolymeric sequences of acidic amino acids – our interest being prompted by the common identification of such repetitive motifs in the primary sequence data abstracted from the native proteins of mineralised tissues [8]. When oligomeric repeats (15mers, 30mers) of aspartic acid or glutamic acid were assayed the habit specific modification of the calcite was observed with the selective stabilisation of {04.4} faces dominating the morphology to produce pseudo-hierarchical crystals preferentially elongated along the [00.1] axis (Figure 2b). The specificity of this interaction suggesting yet again the preferential 'docking' of the oligomer onto one unique subset of lattice planes. More significant perhaps were the observations that (i) the indentation hardness of these peptide-modified crystals (133 kg mm⁻²) increased relative to controls (113 kg mm⁻²), and (ii) the resulting fracture surfaces evidenced a conchoidal profile in contrast to the normal {10.4} cleavage planes of calcite (Figure 2c).

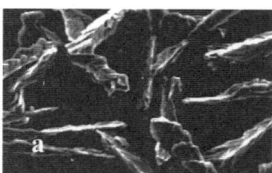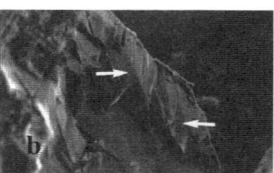

Figure 2. (a) Calcite modified by (Asp)₃₀ expressing {044} faces; (b) fractured surface of crystals in (a) showing conchoidal fracture profile instead of the expected {10.4} cleavage fractures; (c) SEM image of an Echinoderm spine.

From these data one might conjecture that the selective occlusion of the peptide onto the {04.4} planes generates a *nanocomposite* structure. This proposal is further supported by the observation of selective local disordering in the FTIR spectra. The asymmetry of the (v_3) out of plane vibrational mode of the carbonate anion is preferentially increased relative to all other molecular modes of vibration; an observation which is commensurate with the stereoselective occlusion of the protein analogue into the lattice. These studies offer an extension to the earlier work of Addadi *et al.* who evidenced the import of occluded acidic matrix proteins isolated from sea urchin spines in defining the fracture properties of these structures (Figure 2c) [20, 21].

PROGRAMMED TWO DIMENSIONAL TEMPLATES - *CONTROL OF CRYSTAL NUCLEATION*

Studies of one-dimensional templates have focussed attention upon the importance of structural, and stereochemical relationships of host-guest chemistry in the management of additive-mediated crystal *growth* processes. Of equal if not greater interest is the facility for controlling *nucleation* as well as growth. Certainly there are many exemplars of matrix-mediated nucleation in biology. However, as noted in the foregoing discussion, biologically derived mineralisation substrates are difficult to secure in their native form thus, considerable effort has been devoted to the design of ordered two-dimensional synthetic templates capable of controlling nucleation. Langmuir monolayers formed from insoluble amphiphiles at the gas/liquid interface have been used as simple and versatile nucleants for inorganics [23]. In this system the chemical functionality of the amphiphiles and their interfacial organisation can be controlled by the appropriate choice of head group identity and surface pressure [24]. In these and related studies the variation of surfactant headgroups, and thus the presentation of an appropriate stereochemical motif, proved to be instrumental in directing the oriented nucleation of crystals on a specific lattice plane. The inductive capacity of bulk phase surfactants (e.g. *hexagonal, cubic*) has also been exploited in the formation of novel mesoporous organic/inorganic hybrid materials [25-28]. In each case the well-defined macromolecular template presents a highly ordered array of chemical functionality (*analogous to the matrix proteins*). The evolved structure induces mineralisation with the particle size, crystallographic orientation and polymorph selectivity dictated by the chemical identity of the functional groups (e.g. carboxylate, amine, hydroxyl, sulphate, etc.) and the molecular (*spatial, stereochemical*) motifs presented at the interface.

This biomimetic approach was further developed when the utility of chemically functionalised solid/liquid interfaces (*including cross-polymerised Langmuir Blodgett films, covalently self-assembled monolayers and reactive polymer and biopolymer surfaces*) was identified. In this case, applications-oriented research has focused to date upon the near-ambient solution growth of ceramic thin films, especially oxides and hydroxides, on single phase, functionalised self-assembled monolayers (SAMs [29]) [30]. The regio-specific mineralisation of SAMs can be achieved by the selected topographic patterning of the host metal surface with a compatible auxiliary metal prior to the assembly of the organic film [31]. Yet another variant of this approach has been the recent work of Whitesides who has developed a prototype a multicomponent template assembly process which merges lithographic patterning technologies with SAM techniques [31]. In this case the mineralisation template is derived from the co-assembly of alkanethiolate decorated gold surfaces and surfactant hetero-bilayers; the resulting surface can then be mineralised.

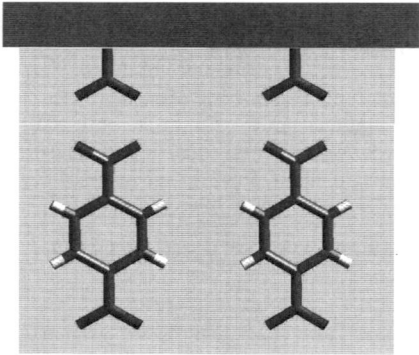

Primary component (I):
COOH functionalised
insoluble surfactant

Crystallisation media doped
with secondary recognition
component (II)

Figure 3. *Schematic detailing design concept adopted for auto-assembling two component templates. The primary component is a carboxylate functionalised insoluble amphiphile. The secondary recognition component is a soluble entity, which will link via metal chelation and/or hydrogen bonding etc. with the primary partner.*

Despite these reported successes the limitation of the experimental design in all cases was the need for external manipulation to direct template assembly. The requirement is clearly for programmed mineralisation templates which *auto assemble* and in doing so create the required structural motif to favour nucleation. The use of photopolymerisation reactions [32,33] has to some degree obviated this need. Yet another development has been the use of functionalised self-assembling amphiphiles which auto-aggregate into hierarchical polymeric systems through hydrogen bonding [34]. More recently this latter concept has been developed further and the utility of programmed *two component self-assembling recognition templates* has been explored.

Self-ordering insoluble amphiphiles may be further functionalised by the co-operative adsorption from the subphase of a secondary soluble component [35]. The secondary recognition units may be selected on the basis of their ability to interact with the insoluble amphiphiles via covalent, electrostatic or H-bonding interactions. One such well-known interaction of organic molecules is the dimerisation of carboxylic acids by metal chelation. In solution, the dimerisation constant for these interactions increases with an increase in alkyl chain length and it is also concentration dependent [36-38]. A template may thus be formed which presents the spatial register dictated by the bulk ordering of the amphiphilic component and is further programmed by the acquired chemical functionality of the soluble recognition unit. To test the feasibility of this approach we have explored the binding of functionalised benzoic acid derivatives (II) to stearic acid monolayers (I). The concept is illustrated schematically in Figure 3 above. These experiments have established that monolayers can be further functionalised by adsorption of a secondary component from the sub-phase. Here the results of applying these templates to a crystallisation assay primed for calcium carbonate precipitation [39] are reported.

In this system polymorph selection is dependent upon the identity of the suphase component. To stabilise calcite, a recognition unit presenting with a carboxylic acid functionality must be used (Figures 4 & 5b). Vaterite stabilisation is achieved with 4-

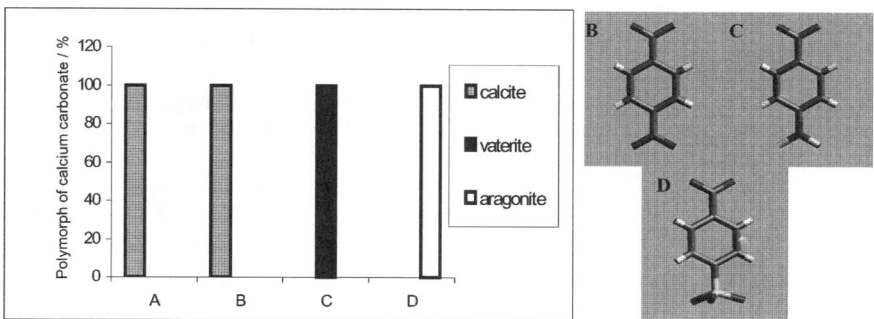

Figure 4. *Graph showing how the additive affects a change in polymorph selectivity.* **A** = *Stearic acid (SA) only,* **B** = *SA with Terephthalic acid,* **C** = *SA with 4-Aminobenzoic acid, and* **D** = *SA with 4-Sulfobenzoic acid*

aminobenzoic acid as the subphase recognition unit (Figures 4 & 5c), whereas aragonite formation requires the adsorption of 4-sulfobenzoic acid to the monolayer (Figures 4 & 5d).

It our contention that polymorph selection will be dependent upon kinetic factors, i.e. charge accumulation, and the developing Ca: CO_3^{2-}, HCO_3^- ratios at the interface rather than specific structural factors. However, it is notable that in all cases the operative second functionality on the benzene ring, must be *meta* to the carboxylate group of the benzoic acid. *Ortho* and *para* substituted units (e.g. 2-Sulfobenzoic acid, 3-Aminobenzoic acid) do not effect any control over crystal nucleation. In addition to the observed polymorphic selection there is clear evidence that all the crystals are preferentially oriented relative to the two component template. In the absence of a secondary recognition component stearic acid monolayers promote the oriented nucleation (<10.0>) of calcite [23] (Figure 5a). In contrast, the adsorption of terepthalic acid favours calcite nucleation on a {01.2} face (Figure 5b), whilst [001] oriented aragonite nucleates with 4-sulphobenzoic acid (Figure 5c) and the adsorption of 4-aminobenzoic acid promotes the nucleation of <11.0> oriented vaterite (Figure 5d).

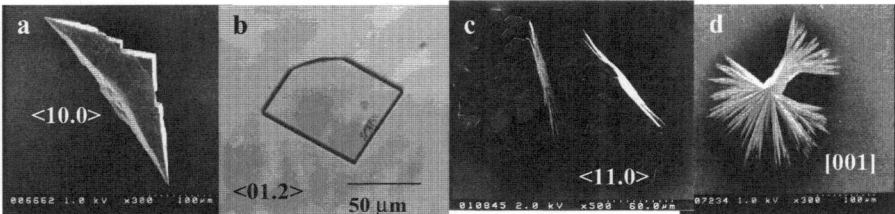

Figure 5. *Micrographs of crystals harvested from two component template assembled in calcium carbonate crystallisation media (pH₀ = 5.8; [Ca²⁺]₀ = 10 mM; T = 25 °C). (a) = Stearic acid (SA) only, (b) = SA with Terephthalic acid, (c) = SA with 4-Aminobenzoic acid, and (d) = SA with 4-Sulfobenzoic acid.*

One might postulate that the role of the template is to lower the activation energy for nucleation. Simple catalysis involving electrostatic interactions at the inorganic-organic interface would be sufficient, but a more complex model dependent upon the template-mediated stabilisation of the inorganic nuclei in a specific transition state by structural and chemical interfacial recognition processes seems more likely. In the present context the observed orientation can be rationalised in terms of a template/crystal match at the organic–inorganic interface defined by both geometric lattice matching and the steric motif of the secondary recognition component (II). The net result of this process is a modification of the reaction dynamics, which in turn promotes oriented nucleation and polymorph selection

THREE DIMENSIONAL TEMPLATES - *CONTROL OF CRYSTAL NUCLEATION & GROWTH*

As noted earlier, one distinguishing feature of many biomolecules associated with mineral deposition is their ability to self–assemble *in vitro* and generate unique supra-molecular architectures. Moreover it is these supra-molecular structures which are credited with being conducive to crystallite formation. The question, which then arises, is whether it is possible to mimic the utility of these structures. The report that some biogenic proteins associated with tissue mineralisation (e.g. ferritin, tuftelins, amelogenins [40,41]) exist as self-ordered spherical arrays, or 'nanostructures', prompted consideration of synthetic analogues capable of mimicking the three dimensional character of these bio-templates.

Dendrimers are three dimensional, highly ordered macromolecules with a branching structure created by a series of iterative chemical changes [42]. The size, shape, topology and surface chemistry of these polymeric species can be regulated in a prescriptive manner by varying the number of iterative synthetic steps (generation number). Recently, we have shown that dendrimers and hyperbranched polymers have some utility in the production of novel inorganic-organic composite materials. Polyimidoamine (PMAM) dendrimers (Figure 6), for example, will modify the habit and polymorphic form of calcium carbonate [43,44]. Indeed, carboxylate functionalised PMAM dendrimers will selectively habit modify calcite and stabilise the metastable calcium carbonate polymorph, vaterite in a generation dependent manner. At a critical dendrimer Generation number (3.5), highly ordered polycrystalline vaterite spheres are

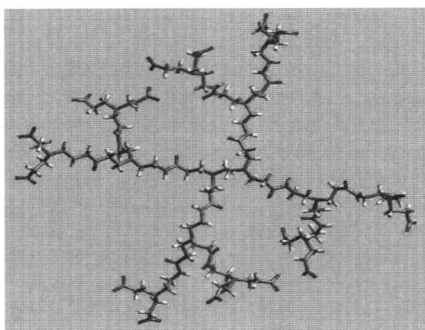

Figure 6. *Computer generated model of carboxylate functionalised PMAM dendrimer Gen 1.5*

favoured over the precipitation of the thermodynamic product, calcite. The importance here of attaining the optimum spherical geometry was recently confirmed by studies with hyperbranched polymers designed to express a similar conformational identity [45].

Here we have investigated further the ability of the PMAM dendrimers to control the nucleation and growth of a *non-polymorphic* system, barium sulphate. Barium sulphate as a biomineral is found in some algae [46] and deep sea species [47]. It has been the subject of many crystals growth, morphological and spectroscopic studies [48-50] largely because of its very low solubility which means that it causes major scaling problems in offshore drilling [51]. The management of these problems is directed towards the prevention of scaling by the addition of inhibitors to retard or block growth processes [51,52]. Given its inherent *plastic* morphology barium sulphate is a valuable tool in fundamental studies of crystal growth phenomena with application to materials chemistry.

In biological, geological and inorganic syntheses, the {001} and {210} faces dominate the morphology (Figure 7f). The addition of dendrimer to the crystallisation media promoted crystal distinctive changes in the habit of the resulting crystals which was strongly generation dependent. Low generation numbers stabilised the dipolar {011} faces, and high generation numbers stabilised the {010, {101} and {410} faces (Figures 7 a & b). Generation 2.5 was unique in catalysing the formation of a monodispersed population of novel spherulites. These spherical aggregates of co-oriented crystallites arise from the repetitive twinning of $BaSO_4$ on the {410} plane; a highly unusual growth mode for this inorganic phase (Figures 7 c, d & e). Analysis reveals that the computed spacing of the carboxylate groups of the Gen 2.5 PMAM dendrimers is an exact geometric match for the lattice co-ordinates of the {410} face with the commensurate template/crystal stereochemistry contributing to polymer adsorption and the molecular recognition events that direct twinning. In contrast, amine functionalised dendrimers do not effect any change in the crystals (Figure 7f). Work is continuing to define more specifically the association of the dendrimer with the organic phase and the role of this supramolecular construct, and related polymers, in the formation of novel inorganic-organic composites.

CONCLUSION

The use of biologically derived matrices to control the nucleation and growth of inorganic materials has attracted considerable interest given the unique form and functional optimisation of the biominerals. Indeed, the pivotal role of defined bio-polymeric matrices in controlling bio-crystallisation strategies is undisputed but remains essentially undefined. Unfortunately, in many cases the native configuration of the matrix is frequently compromised during the isolation procedures and evidences little control of mineralisation in *in vitro* assays. Here we have investigated the utility of systematically changing the chemical functionality and interfacial organisation of a series of synthetic templates whose design was inspired by consideration of the chemical organisation of their bio-polymeric analogues. These studies have indicated that the selective "host-guest" principles of molecular recognition and molecular self-organisation operate to control certain aspects of the nucleation, growth and stabilisation of inorganic particles. The size, crystallographic orientation and growth of a mineral phase can be controlled and given a template of appropriate design used to construct of organised inorganic/organic composites. Clearly then, there is much potential in adopting biomimetic solid state chemistry as part of an innovative approach to the tailored fabrication of crystalline materials.

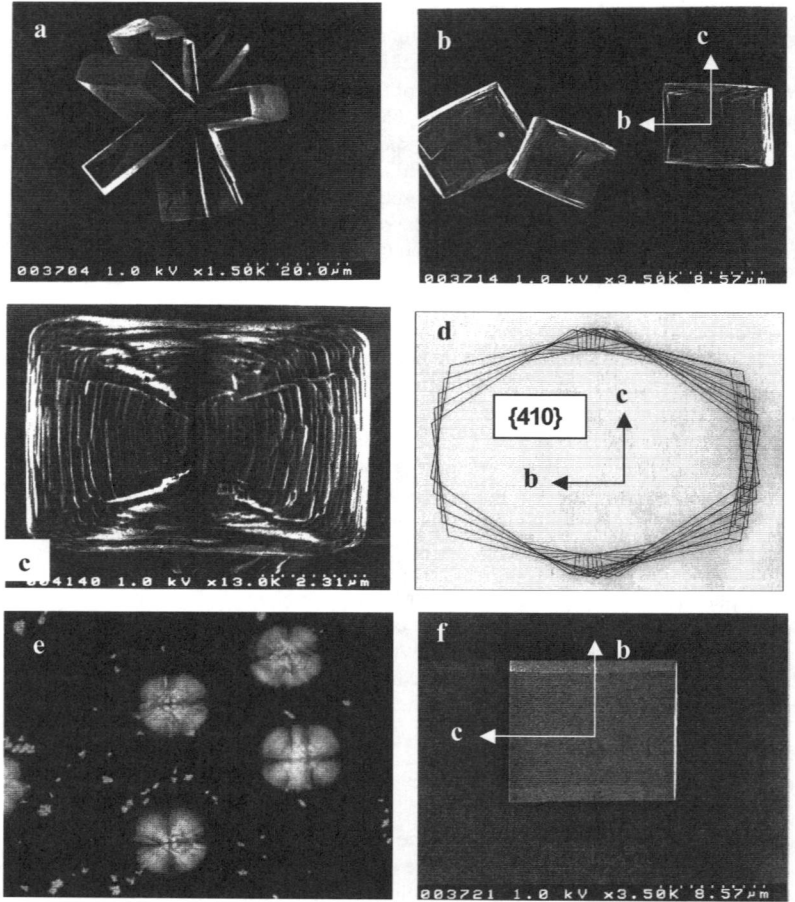

Figure 8. (a) FESEM of barium sulphate crystal grown with CO_2H terminated PMAM Dendrimer, Gen 0.5; (b) FESEM of barium sulphate crystal grown with CO_2H terminated PMAM Dendrimer, Gen 1.5; (c) FESEM of barium sulphate crystal grown with CO_2H terminated PMAM Dendrimer, Gen 2.5; (d) Schematic detailing crystallographic features of crystals depicted in (c); (d) Optical image of crystals in (c) secured using polarised light microscopy. The Brewster Interference pattern is indicative of an anistropic crystalline aggregate; (f) FESEM of barium sulphate crystal grown with NH_2 terminated PMAM Dendrimer, Gen 2.0.

REFERENCES

1. S. Weiner and L Addadi, *J Mat. Chem.* **7** 689-702 (1997)
2. B. C. Jeong, C Hawes, K. M. Bonthrone and L. E. Macaskie, *Microbiology* **143**, 2497-2507 (1997)
3. K.O. Konhauser, *FEMS Micobiology Reviews*, **20**, 315-326 (1996); T Klaus, *Proc Natl Acad Sci*, **96**, 13611-13613 (1999)
4. H. Lowenstam and S. Weiner, 'On Biomineralisation' (Oxford University Press 1989)
5. L. Addadi and S. Weiner, *Angew Chemie. Intl. Edn. Engl*, **31**, 153-169 (1992)
6. S. Mann, *Nature* **365**, 499-505 (1993)
7. M. S. Fernanedez, M. Araya and J. L. Arias, *Matrix Biology*, **16**, 13-20 (1996)
8. S. Albeck, L. Addadi and S. Weiner, *Conn. Tiss. Res*, **34-35**, 419-424 (1996); M.E. Marsh and D. P. Dickinson, *Protoplasma*, **199**, 9-17 (1998)
9. W.T Butler, H.H. Ritchie and A. L. J. J. Bronckers, *Ciba Foundation Symposium* **205** 107-117 (1998); A. George, L. Bannon, B. Sabsay, J. W. Dillon, J. Mallone, A. Veis, N. A. Jenkins, D. J. Gilbert and N. G. Copeland, *J. Biol. Chem*, **271**,32869-32873 (1996); K. L. Hirst, K. O'Connor, M. F. Young, and M. J. Dixon, *J Dent. Res*, **76**, 754-760 (1996); M. l. Paine and M. L. Snead, *J. Bone Min Res*, **12**, 221-227 (1997)
10. A. K. Powell, *Structure and Bonding*, **88**, 1-38 (1996); N. Kroger, R. Deutzmann, M. Sumper, *Science*, **286**, 1129-1130 (1999)
11. J. W. Dana and E. S. Dana, *The System of Mineralogy* (Wiley, New York , 1951); R. J. Reeder, *Reviews in Mineralogy*, **11**, 1-47 (1983)
12. K. L. Hirst, D. Simons, J. Feng, H. Aplin, M. J. Dixon and M. McDougall, *Genomics*, **42**, 38-45 (1998)
13. P. A. Bianconi, J. Lin and A. R. Strezelecki, *Nature*, **349**, 315-317 (1991)
14. B. Brisdon, B. R. Heywood, A. Hodson, S. Mann and K Wong, *Adv. Mat*, **5**, 449-512 (1993)
15. J. M. Marentette, J. Norwig, E. Stockelmann, W. H. Meyer and G. Wegner, *Adv Mat*, **9**, 647-651 (1997)
16. N. Puricha and L. C. Erway *Dev. Biol*, **27**, 395-405 (1975); A. Bigi, E. Foresti, M. Gandolfi, M Gazzano and N. Roveri, *J Inorg Chem*, **58**, 49-58 (1995); T. Kitijama, M. Tomita, C.E. Killian, K. Akasaka and F. H. Wilt, *Dev. Growth Diff*, **38**, 687-695 (1996)
17. J. Didymus, S. Mann, N. Sanderson and B. R. Heywood *J Chem Soc, Farad Trans* **86**, 1873-1880 (1989)
18. D. Carlstrom, *Biol Bull*, **125**, 441-463 (1963)
19. K. G. Pote and M. Ross, *Comp Biochem Physiol* **98B**, 287-295 (1991)
20. A. Berman, L. Adddadi and S. Weiner *Nature*, **331**, 546-548 (1988)
21. J. Aizenberg, J. Hanson, T F. Koetlze, L. Leiserowitz, S. Weiner and L. Addadi, *Chem Eur J*, **7**, 414-422 (1995)
22. E. M. Landau, M. Levanon, L. Leiserowitz, M. Lahav, and J. Sagiv, *Nature*, **318**, 353-356 (1985)
23. B. R. Heywood and S. Mann, *Adv. Mat*, **6**, 9-20 (1994)
24. G. Gaines, *Insoluble Monolayers at Liquid-Gas Iinterfaces*, (Wiley Interscience, New York, 1966)

25. J. Brinker, *Nature* **394**, 256 (1998)
26. S. Oliver, A. Kuperman, N. Coombs, A. Lough and G. Ozin *Nature* **382**, 589 (1996)
27. K. Tanav and P. Pinnavaia, *Science* **267**, 865-868 (1995)
28. H. Yang, N. Coombs and G. Ozin, *Nature* **386**, 692-695 (1997)
29. A. Ulman, *Introduction of Thin Organic Films; From Langmuir-Blodgett to Self-Assembly* (Academic Press, Boston, 1991)
30. B. Bunker, *Science*, **264**, 48-55 (1994)
31. J. Aizenberg, A. J. Black, G. M. Whitesides, *Nature* **394***, 868-871* (1998)
32. A. Berman, D. J. Ahn, A Lio, M. Salmeron, A Reichart and D Charych, *Science* **269**, 515-518 (1995)
33. S. M. D'Souza, C. Alexander, S. W. Carr, A. M Waller, M.J. Whitcombe and E. N. Wulfson, *Nature* **398**, 312-316 (1999)
34. P. Fallon, B. R. Heywood, M. Mascal and S. J. Williams, *Angew Chemie*, in press (2000)
35. H. Ringsdorf, B. Scharlb and J. Venzmer, *Angew Chenie (Intl Ed Engl)* **27**, 113-158 (1988)
36. N. Tanaka, H Kitano, N. Ise, *J Phys Chem*, **4**, 6290-6292 (1990)
37. V. Daon, R. Koppee and P. H. Kasai, *J Am Chem Soc*, **119**, 9810-9815 (1997)
38. A. Katalsky, H. Gisenberg and S. Lifson, *J Am Chem Soc*, **73**, 5889-5890 (1951)
39. Y. Kitano, *Bull Chem Soc Jap*, **35**, 1973-1980 (1962)
40. A. Fincham and J. P. Summer, *Ciba Foundation Symposium*, **205**, 118-134 (1997)
41. M. L. Piane, D. Deutsch and M. L. Snead, *Conn Tiss Res*, **34/35**, 211(1996)
42. D. D. Tomalia, *Angew. Chem. Int. Ed. Eng* **29**, 138 (1990)
43. B. R. Heywood, C. S. German, S J Hill and S.J. Williams, *J Mat Chem*, in press (2000)
44. B. R Heywood S.J. Hill and S.J. Williams, *Adv Mat*, in press (2000)
45. W.J. Feast, B.R.Heywood, L Hobson, S.J.Williams, *J Mat Chem*, in press (2000)
46. D. R. Kreger and H. Boere, *Acta Bot Neerl.*, **18**, 143 (1969)
47. A. J. Gooday and J. A. Nott, *J Mar Biol Ass, UK*, **62**, 595 (1982)
48. M. C. Van der Leeden, M. C. Reedijk, and G. M. van Rosmalen, *Estudios Geol.*, **38**, 279 (1982)
49. M. C. Van der Leeden and G. M van Rosmalen, *SPE Production Engineering*, **70** (1990)
50. B. R. Heywood and S. Mann, *Langmuir* **8**, 1492-1494 (1992)
51. R. J. Benton, *Faraday Discussions*, **95**, 281-287 (1993)
52. R. J. Davey, S. Black, L. Bromley L. Cottier, B. Dobbs and L. Rout *J Chem Soc, Farad Trans*, **87**, 3409-3413 (1991)

Mat. Res. Soc. Symp. Vol. 620 © 2000 Materials Research Society

Amelogenin Nanospheres Modulate Crystal Habit of Octacalcium Phosphate and Hydroxyapatite Crystals in *In Vitro* Model Systems

[1]Moradian-Oldak J., [1]Wen H.B., [1]Fincham, A.G. and [2]Iijima M.
[1]Center for Craniofacial Molecular Biology, School of Dentistry, University of Southern California, 2250 Alcazar St. LA,CA 90033. [2]Asahi University School of Dentistry, Dental Material and Technology 1851-1 Hozumi-cho, Motosu-gun, Gifu 501-02, Japan

ABSTRACT

This paper is a short review of recent studies, which were undertaken to investigate interactions of amelogenin with octacalcium phosphate (OCP), and apatite. OCP crystals were grown using two independent experimental systems; (a) in a 10% gelatin gel, containing 0-2% amelogenin, where the crystals were formed in a double-diffusion chamber, and (b) in a 10% pure amelogenin gel, where crystal growth took place in between a cation-selective and a dialysis membrane. Apatite crystals were grown from a supersaturated calcifying solution on a bioactive glass in the absence (SCS_B) and the presence of amelogenin (SCS_{rM179}). It was found that OCP crystals formed in 10% gelatin gel containing 1-2% amelogenin were longer (3-5 times larger in aspect ratio) than the OCP crystals formed in 10% gelatin without amelogenin. A profound effect was that found in the cation selective membrane system when 10% amelogenin inhibited the growth morphology in a specific manner. Affected crystals had a length to width ratio twice larger than that of control crystals while the width to thickness ratio was about 1/12 of that of the control crystals. Amelogenin promoted the formation of bundles of lengthwise apatite crystals, which were all oriented parallel to their c axes when grown on SCS_{rM179}. It was found that individual apatite crystals within those bundles adopted an elongated, curved shape. The data presented here suggest that amelogenin nanospheres modulate the growth morphology of apatite and OCP crystals and indicate significant functional roles for amelogenin proteins during the *in vivo* oriented growth of enamel crystallites.

INTRODUCTION

The calcium carbonated hydroxyapatite crystals that comprise the mineral phase of mature enamel are the longest biogenic apatites in nature. The formation of enamel apatite crystals takes place in an amelogenin -rich organic matrix whose components are continuously secreted, self-assembled and eventually processed to almost complete degradation. It is now accepted that amelogenin "nanospheres", the principal protein component of developing tooth enamel constitute the basic structural units of the enamel extracellular matrix structural framework [1] (figure 1). Recent studies have indicated that amelogenin "nanospheres" may play critical roles during the initiation and growth of enamel crystallites by providing the dynamic structural framework (or network) within which the crystals grow. The 20nm amelogenin nanosphres were demonstrated to be formed as the result of assembly of approximately 100 protein molecules [2] and they have been postulated: a) to provide a protective envelope around the tiny apatite crystallites at the early stage of enamel formation preventing their lateral fusion, b) to limit growth of the crystals in the direction perpendicular to a and b axis resulting preferential c-axial growth through the full thickness of the secretory matrix. Although amelogenin proteins are hydrophobic in nature, the hydrophilic domain at the carboxy-terminal region provides the nanospheres with charged surfaces, an interface through which the protein can interact with ionic crystals. The apatite binding properties of amelogenin proteins have been investigated in terms of adsorption parameters and kinetic measure [3]. We have recently reported that the recombinant mouse amelogenin rM179 slightly inhibited the growth of apatite crystals from seeded solution and showed an "adherence" effect on apatite crystals, causing their aggregation [4]. These interactions were suggested to occur through the hydrophilic carboxy-terminal region by the "polymer-bridging mechanism", presumably resulting from the supramolecular self-assembly of amelogenin nanospheres [4]. Although the above studies have indicated some interactions of amelogenin proteins with growing apatite crystals, details on the specificity of these interactions were lacking. With the recent advance in the structural biology of amelogenin protein, namely identification and characterization of the nanospheres [1], current studies focus on the influence of amelogenin nanospheres on the morphology of octacalcium phosphate and apatite crystals. The overall goal was to examine the hypothesis that amelogenin nanospheres provide the appropriate microenvironment for the oriented and lengthwise growth of enamel apatite crystallites. We plan to explore general principles governing interactions between amelogenin nanospheres, as organic matrix framework and apatite crystals. Investigating such molecular mechanisms will advance understanding of basic general principles in other biomineralizing tissues, and will contribute to the knowledge for developing novel biomimetic materials.

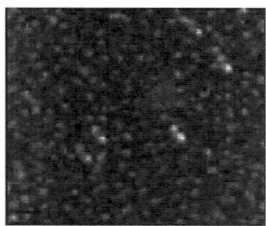

Figure 1. Tapping mode atomic force microscope image (1x1μ field) in air (Multimode Nanoscope III, Digital Instruments) of the "25kDa" amelogenin extracted from developing porcine molars, isolated and purified by HPLC techniques as previously described [5]. (Bar=100nm). The sample was prepared according to Moradian-Oldak et al. [6]. Scale on the right presents height of the image.

EXPERIMENTAL STRATEGIES

Octacalcium phosphate (OCP) crystals were used as a primary model crystal system for the study of matrix mediated enamel biomineralization and they are appropriate because: a) there is evidence supporting the hypothesis that the ribbon-like enamel crystallites grow initially as OCP [7], b) apatite crystals grown in *vitro* under semi-physiological conditions are small and fragile, c) apatite crystals are non-stoichiometric in nature and there is inconsistency between crystal structure (hexagonal) and crystal morphology (plate-like morphology), factors that complicate the study of amelogenin-apatite interactions. A recombinant mouse amelogenin expressed in *e-coli* and purified as previously described [8] and amelogenins extracted from developing porcine molars were used as protein sources. Two independent *in vitro* experimental systems were used to investigate amelogenin interactions with octacalcium phosphate crystals while the effect of amelogenin on apatite crystal growth was examined using the bioactive glass surfaces.

1. A double diffusion device was used for growth of octacalcium phosphate in 10% gelatin gel loaded with different concentrations of porcine amelogenins [9]. Buffered calcium and

phosphate solutions (pH 7.4) were circulated from the opposite ends of the gel, at room temperature for 7 days at a continuous rate through this device. The mineral discs precipitated in the gels were characterized by TEM, SEM and XRD.

2. A dual membrane system was developed allowing oriented growth of OCP crystals between a cation selective membrane and a dialysis membrane when the crystals grew on the cation selective membrane [10-12]. Calcium ions (pH 6.5) were supplied into the phosphate solution (pH 6.5) through the cation selective membrane into the space between the two membranes. The system was designed to examine the growth of OCP crystals in pure recombinant mouse amelogenin gel, rM179 (100mg/ml). Albumin and gelatin were used as control proteins.

3. Bioactive glass, the Bioglass® 45s5 type, a bioactive material or calcium phosphate inducible was used as a substrate for growth of apatite crystals from a supersaturated calcifying solution [13]. Apatite crystals were grown by incubation of the bioglass in phosphate buffer saline solution for one week at 37 °C and immersion in a supersaturated calcifying solution for 72 hours. The silanol groups on the hydroxylated Bioglass ® surface promoted the oriented nucleation of apatite crystals with their c axis perpendicular to the surface of Bioglass ®. This epitaxial oriented nucleation on the bioactive surface has been confirmed by the strong 002 diffraction of apatite crystals precipitated after PBS incubation. The effect of the recombinant amelogenin rM179 on the growth of apatite crystals was investigated by adding the protein to the calcifying solution.

RESULTS AND DISCUSSION

Octacalcium phosphate in gelatin gel in the absence and presence of amelogenin
The use of gelatin gel in a double diffusion device allowed examination of the effect of amelogenin concentration on the growth morphology of octacalcium phosphate crystals. The criteria for determining effect of protein to crystal morphology was based on measuring aspect ratio, namely length to width ratio. X-ray diffraction of the mineral formed in the gel has confirmed the presence of OCP [9]. Figure 2 shows scanning electron micrographs of OCP crystals grown in 10% gelatin gel in the absence (A), and the presence (B) of 1% (10mg/ml) amelogenin proteins extracted from developing porcine enamel matrix. OCP crystals grown in 10% gelatin gel were found to be dispersed plates, mostly 0.5-3 micron long and 0.1-0.5 micron across (figure 2A). Remarkably, the crystals deposited in the amelogenin containing gel appear to have an elongated shape (figure 2B). Selected area electron diffraction of isolated OCP crystals has confirmed the elongation to be parallel to the c axis. In some cases, bundles of crystals were observed were the crystals within those bundles appeared to be oriented parallel to one another in their c axis. Crystal growth experiment at different concentration of amelogenin ranging from (0-20 mg/ml) showed increase in aspect ratio (L/W) of OCP crystals as a function of amelogenin concentration in gelatin gel. This dose dependent behavior may indicate that amelogenin nanospheres selectively interact with the lateral phases of OCP inhibiting the growth in a- or b- axial direction.

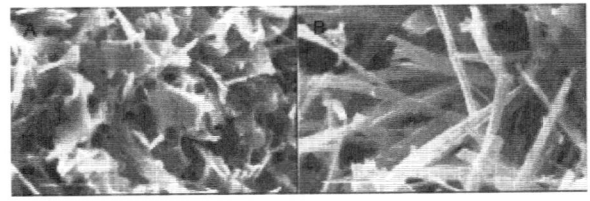

Figure 2. *Scanning electron micrographs of OCP crystals grown in 10% gelatin gel by double diffusion technique; (A) in the absence, (B) in the presence of 1% (10mg/ml) amelogenin.(Bar=2μm)*

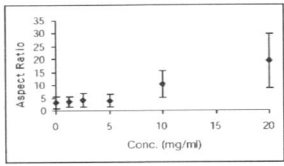

Figure 3.) *Aspect ratio (L/W) of OCP crystals grown in 10% gelatin gel as a function of amelogenin concentration*

Octacalcium phosphate in amelogenin gel. In a model system of enamel formation Iijima and Moriwaki (1999)[10] have used the cation selective membrane to control the ionic diffusion and have demonstrated that the one-directional supply of Ca^{+2} ions through the membrane promoted the lengthwise and oriented growth of octacalcium phosphate crystals. We used this system to examine the effect of amelogenin gel on the growth of OCP crystals with compare to those of albumin and gelatin. X-ray diffraction patterns of OCP crystals indicated that in general addition of proteins caused inhibition of crystal growth when the degree of inhibition was in the order of gelatin<rM179<albumin. OCP crystallized in its characteristic ribbon-like shape in the presence of albumin and gelatin. The effect of amelogenin proteins on the morphology of OCP crystals was determined by estimating the length/width and width/thickness ratio. Figure 4 is the scanning electron micrograph of OCP crystals grown in the absence (figure 4A) and the presence (figure 4B) of 10% recombinant amelogenin rM179. In the presence of amelogenin rod-like and prism-like crystals were formed predominantly while the control crystals showed the typical ribbon-like morphology. In general, smaller crystals were formed in the presence of amelogenin indicative of some inhibitory activity. Comparison of crystal aspect ratio between control and amelogenin gel suggested this inhibitory activity to be specific. Aspect ratio (L/W) of the crystals grown in amelogenin gel was twice larger than that of control. Additional interesting observation was that the W/T ratio decreased (1/12) when the crystals grew in amelogenin. These observations suggest that amelogenin protein selectively inhibited the growth in the *b* axis direction.

Apatite nucleated on bioactive glass. As an alternative model for the seeded crystal growth, we used the Bioglass ® 45s5 type, a system that results in the oriented growth of apatite crystals. Figure 5 presents the scanning electron micrographs of apatite crystals grown from a supersaturated calcifying solution in the absence (SCS_B), (figure 5A) and the presence (SCS_{rM179}) (figure 5B) of 50μg/ml amelogenin rM179. The calcium phosphate layers formed on the treated

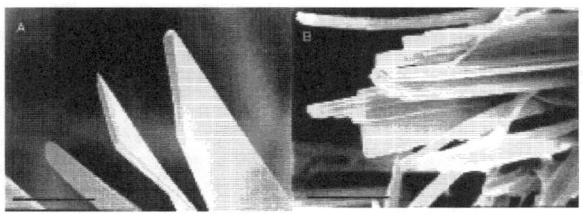

Figure 4. SEM micrographs of OCP crystals grown in the cation selective membrane system; (A) control, (B) 10% (100mg/ml) recombinant amelogenin rM179. (Bar=1μm)

samples were characterized by X-ray diffraction and Fourier transmission infrared spectrometry to be apatite. As seen in the figure plated-shaped crystal bundles of 50 nm thick and 300-600 across were observed on the control (figure 5A) when the mineral bundles deposited on the SCS_{rM179}-immeresed samples were thicker adopting an elongated curved shape. The TEM microghraph (inset in figure) of crystal bundles scraped from the surface of the Bioglass showed that the crystals formed on the SCS_{rM179} composed of bundles of elongated individual crystals (15-20 nm thick) oriented parallel to each other. Immersion of SCS_B in albumin containing calcifying solution (SCS_{BSA}) resulted in the overall reduction of crystal sizes indicating a general inhibitory activity [13]. Comparison of X-ray diffraction patterns of crystals formed on the surface of bioglasses in the presence of amelogenin and albumin showed the 002 diffraction of apatite to be stronger in SCS_{rM179} than in the case of SCS_{BSA} indicating c axial elongation of apatite crystals in the amelogenin-containing sample. It was therefore suggested that amelogenin inhibited the lateral growth of apatite crystals allowing their prefrential growth in the c axial direction.

CONCLUSIONS

Amelogenin proteins isolated from developing enamel extracellular matrix are capable to interact with octacalcium phosphate crystals and affect the length to width ratio resulting in their elongation in the *c* axis. This effect was found to be dose dependent indicating selectivity in interaction of the protein with the lateral phases (phase parallel to the *c* axis) of octacalcium phosphate crystals. In the cation selective membrane system recombinant amelogenin decreased the overall size of octacalcium phosphate crystals. These crystals grown in 10% amelogenin gel appeared to be more elongated when compared to those grown in 10% gelatin or albumin and were found to pre dominantly be affected in their *b* axis direction. Recombinant amelogenin rM179 added to a supersaturated calcifying solution caused adherence of oriented apatite crystals nucleated on the surface of a Bioglass ®. In this system amelogenin nanospheres also resulted in the elongation of apatite crystals in their *c*- axial direction. Collectively, these observations support the hypothesis that amelogenin proteins are directly involved in the elongation of enamel apatite crystals during the early stage of enamel development. The molecular mechanisms of these interactions still need to be elucidated.

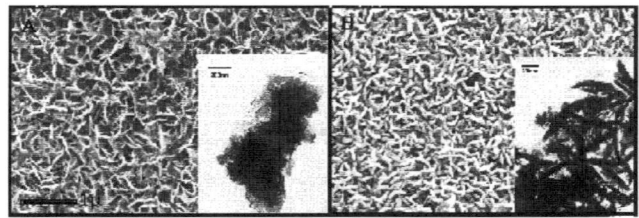

Figure 5. SEM micrographs of apatite crystal bundles grown on the bioglass surface; (A) in the absence and (B) in the presence of recombinant amelogenin rM179. Insets are the TEM micro graphs of crystals scraped from the surface. Note that in the presence of amelogenin apatite crystals adopted an elongated curved shape. (Bar in inset=200nm)

ACKNOWLEDGMENTS. This work was supported by NIH-NIDR research grants DE-02848 and DE-12350. We thank US Biomaterials Corporation (USA) and Asahi Glass Co. (Japan) for providing the Bioglass ® samples and the cation selective membrane.

REFERENCES

1. A.G Fincham , J. Moradian-Oldak, and J.P. Simmer, (1999) *J Struct. Biol.* **122**, 320-327.
2. J. Moradian-Oldak, J.P. Simmer, E.C. Lau, P. E Sarte, H.C. Slavkin, and A.G. Fincham, (1994), *Biopolymers* **34**, 1339-1347.
3. T. Aoba, M. Fukae, T. Tanabe, M. Shimizu, and E.C. Moreno (1987) Adv. Dent. Res. **1**, 252-260.
4. J. Moradian-Oldak, J. Tan, and A.G. Fincham, (1998) *Biopolymers,* **46**, 225-238.
5. J. Tan, W. Leung, J. Moradian-Oldak, M. Zeichner-David, and A.G. Fincham, (1998) *J. Dent. Res.* , **77**, 1388-1396.
6. J. Moradian-Oldak, M.L. Paine, Y.P. Lei, A.G. Fincham, and M.L. Snead (2000) *J. Struct. Biol.* (in press)
7. Y. Miake, S. Shimoda, M. Fukae, T. Aoba 91993) Calcif. Tissue Int. **53**, 249-256.
8. J.P. Simmer, E.C. Lau, C.C Hu, P. Bringas, V. Santos,. T. Aoba. M. Lacey, D. Nelson, M. Zeichner-David, L. S. H. Malcom, Slavkin, and A.G Fincham. (1994) *Calcif. Tissue Int.* 54, 312-319.
9. H.B. Wen, J. Moradian-Oldak, and A.G. Fincham, Amelogenin incorporation into gelatin gels modulates crystal habit of octacalcium phosphate crystals. (in preparation)
10. M.Iijima and Moriwaki (1999) *J Japanese Association for Crystal Growth*, **26**, 175-183.
11. M. Iijima, Y. Moriwaki, T. Takagi, J. Moradian-Oldak, A.G. Fincham (2000) "47th Annual Meeting of JADR", May 1999, J. Dental Res. , abstract No.108 (in press)
12. M. Iijima, Y. Moriwaki, T. Takagi, H.B. Wen, A.G. Fincham, J. Moradian-Oldak (manuscript in preparation)
13. H.B. Wen, J. Moradian-Oldak, and A.G. Fincham, (1999) *Biomaterials* , 20, 1717-1725.

Growth of Organic Films and
Supramolecular Solids

Mat. Res. Soc. Symp. Vol. 620 © 2000 Materials Research Society

Preferential dissolution along misoriented boundaries in heterogenite

R. Lee Penn[1,2], Alan T. Stone[2], and David R. Veblen[1]
[1]Department of Earth and Planetary Sciences
[2]Department of Geography and Environmental Engineering
Johns Hopkins University, Baltimore, MD 21218, U.S.A.

ABSTRACT

High-Resolution Transmission Electron Microscopy (HRTEM) results show a strong crystal-chemical and defect dependence on the mode of dissolution of synthetic heterogenite (CoOOH) particles. As-synthesized heterogenite particles are micron-size plates (aspect ratio ~ 1/30) constructed of crystallographically oriented ~ 3-nm primary particles or are single ~ 21-nm unattached heterogenite platelets (aspect ratio ~1/7). Reductive dissolution, using hydroquinone, was examined in order to evaluate morphology evolution as a function of reductant concentration. Two end-member modes of dissolution were observed: 1) non-specific dissolution of macroparticles and 2) preferential dissolution along misoriented boundaries. In the case of non-specific dissolution, average macrocrystal size and morphology are not altered as building block crystals are consumed. The result is web-like particles with similar breadth and shape as undissolved particles. Preferential dissolution involves the formation of channels or holes along boundaries of angular misorientation. Such boundaries involve only a few degrees of tilt, but dissolution occurs almost exclusively at such sites. Energy-Filtered TEM thickness maps show that the thickness of surrounding material is not significantly different from that of undissolved particles. Finally, natural heterogenite from Goodsprings, Nevada, shows morphology and microstructure similar to those of this synthetic heterogenite.

INTRODUCTION

Central to understanding the geochemical cycling of the elements is elucidating the processes that liberate chemical species and allow them to move. Dissolution is one such process and can liberate chemical species by producing either aqueous molecular complexes or nanocrystalline particles that become important components of mobile fluids in our environment. Understanding growth and dissolution processes requires a holistic characterization of the phases involved. Coupling high-resolution transmission electron microscopy (HRTEM) with wet chemical methods allows direct correlation of atomic-scale changes in dissolving materials with changes in solution and surface chemistry.

Dissolution experiments utilizing particles with different types of surface area have the potential to provide the link between understanding reactive surface area and the chemical behavior of a mineral surface. This work's focus involves dissolution experiments utilizing heterogenite (CoOOH) particles formed by oriented assembly. This mineral belongs to a class of minerals, the oxyhydroxides, that are important because they comprise a large proportion of the reactive surface area in the environment (e.g., in soils and as mineral weathering products [2]). In addition, ^{60}Co is a radionuclide and an appreciable component in radioactive wastes [3].

Co(II) is many orders of magnitude more soluble than Co(III). Thus, compounds capable of reducing Co(III) to Co(II) can bring about the *reductive dissolution* of Co(III) bearing minerals such as heterogenite. Hydroquinone (QH$_2$) has been found to dissolve heterogenite solely via this mechanism (eq. 1), yielding stoichiometric amounts of the product p-benzoquinone (Q) [4]. This

$$2Co^{(III)}OOH(s) + QH_2 + 4H^+ = 2Co^{2+} + Q + 4H_2O \qquad eq. \ 1$$

work complements these previous results by coupling HRTEM examination with wet chemical methods.

EXPERIMENTAL

All suspensions and solutions were prepared using 18 MΩ•cm resistivity water (DDW, Millipore Corp.). Glassware and plasticware were soaked in 5 M HNO_3, rinsed several times with distilled, deionized water, and rinsed several times with DDW prior to use. All chemicals were reagent grade and used without further purification: hydroquinone (99+%, Aldrich), benzo-quinone (99+%, Aldrich), 30 % hydrogen peroxide solution (< 1ppm Tin as a stabilizer, EM Science), NaOH pellets (Aldrich), 6 % NaOCl solution (VWR), and $CoCl_2$•$6H_2O$ (Baker Analyzed).

Preparation of Heterogenite particles:

Heterogenite particles were prepared by first producing a Co(II)-bearing precipitate followed by oxidation using NaOCl. Starting solutions of 2.9 mM NaOH (3.0L) and 1.2 mM $CoCl_2$ (3.0 L) were prepared, heated to 80°C using a constant-temperature water bath, and sparged using Ar overnight. The NaOH solution was delivered to the $CoCl_2$ solution vessel using a peristaltic pump with a flow rate of 40 mL/min. A pale pink suspension resulted. The reaction vessel was hand-rocked to ensure mixing and maintained at 80°C. Eighteen minutes after the NaOH addition was complete, 1.0 L 17.5 mM NaOCl was quickly added; the suspension immediately turned brown. Five hours later, the bottle containing the suspension was removed from the 80°C water bath and placed in a room-temperature water bath.

The total Co concentration, as heterogenite particles, was determined by dissolving the particles using hydrochloric acid and analyzing the resulting solution using flame atomic absorption spectroscopy (AAS). The suspensions contained 500μM Co as heterogenite particles. Synthesis of heterogenite was verified using X-ray diffraction analysis. A small volume of suspension was centrifuged and resuspended using purified water in order to prepare samples for TEM examination.

Dissolution Experiments:

Suspensions of constant Co concentration (0.4 mmol/L as CoOOH) were prepared using an acetate buffer (10 mM, pH ~ 4.6) and varying concentrations of hydroquinone. Suspensions were allowed to react, with constant shaking, for 24 hours at room temperature. Suspension pH consistently rose approximately 1/10th of a pH unit. Particles were recovered by filtering suspensions using 0.1 μm track-etch membranes (Nuclepore Corp.). Particles were resuspended in DDW and one drop of each suspension was placed on a holey carbon coated TEM grid and allowed to dry.

Filtrate was analyzed for hydroquinone, benzoquinone, and total dissolved Co. Hydroquinone and benzoquinone concentrations were determined using a reverse-phase HPLC (Waters Corp.) equipped with a variable-wavelength UV detector, a μBondapak ™ C18 guard column, a μBondapak ™ C18 3.9 x 300 mm column with a 10-μm packing size, and an autosampler. Eluent flow, using two pumps, was adjusted to deliver ~17% methanol (HPLC grade) and ~83% 10 mM acetic acid (filtered and degassed) at a total flow rate of 1mL/min. Each injection volume was 50 μL. Total dissolved Co was determined using AAS.

Electron Microscopy of partially dissolved particles:

All pre- and post-dissolution materials were examined using a Philips CM300FEG TEM equipped with an Oxford light-element Energy Dispersive X-ray (EDX) system, with ESVision software, for compositional analysis and a Gatan Imaging Filter (GIF) for energy filtered imaging. All TEM images were collected using a Gatan CCD and Digital Micrograph software. Filtered TEM images in this work are formed using "zero-loss" electrons. Inelastic scattering events, such as inner-shell ionization events, are excluded by use of a magnetic prism, a part of the GIF system. The exclusion of inelastic scattering improves image resolution and contrast [5].

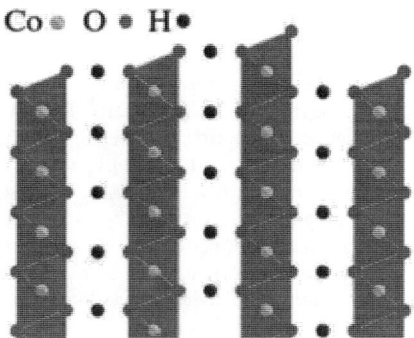

Figure 1: Schematic diagram of the heterogenite crystal structure.

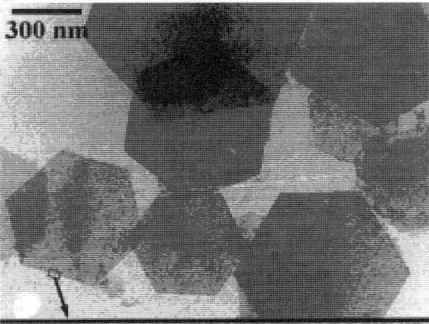

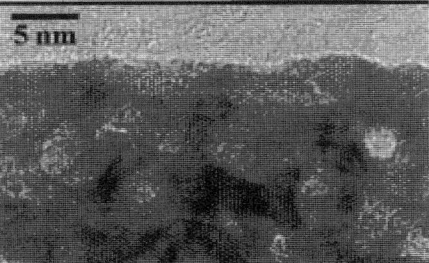

Figure 2: Zero-loss TEM images of heterogenite plates. The lower image is a high-resolution im-age of a small region from the top image.

RESULTS AND DISCUSSION

The heterogenite crystal structure:

The heterogenite crystal structure has hexagonal symmetry and can be broadly described as sheets of edge-sharing Co octahedra stacked along the c-axis (fig. 1). Interlayer protons are situated between oxygen atoms from adjacent layers, one proton for each pair of oxygen anions. The stacking vector, here defined as the angle between [100] and a vector connecting Co cations in adjacent sheets, is inclined with respect to **c** by 10.5°. One unit cell is 3 octahedral layers thick along **c** [6].

Characterization of as-synthesized particles:

The particles are hexagonal plates with an average width of 0.7 μm and an aspect ratio of ~1/30. HRTEM results reveal these plates to consist of networked and oriented ~3-nm nanocrystals, which can be thought of as primary building block particles. Figure 2 shows a zero-loss HRTEM micrograph of a typical heterogenite particle. A low-magnification zero-loss image shows the average morphology of the plates (upper), and a high-resolution zero-loss image (lower) shows distinct boundaries between pores and primary building blocks. While boundaries between perfectly aligned nanocrystals are effectively eliminated when oriented assembly occurs, pores and misoriented boundaries are distinct and amenable to imaging.

Figure 3 shows a zero-loss HRTEM image of a misoriented boundary between primary heterogenite nanocrystals. Grey lines serve to highlight the angular mismatch across the boundary. That misoriented boundaries are incorporated into the larger plate suggests an attachment process rather than growth by atomistic additions, since the driving force to perfectly incorporate single

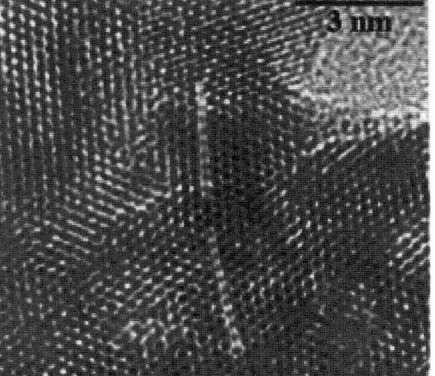

Figure 3: Zero-loss HRTEM image showing a misoriented boundary between primary building block heterogenite nanocrystals.

Co(III) cations is expected to be quite high. Atomic-scale surface roughness, which can take the form of steps or islands on a crystal surface, can result in imperfect attachment, giving rise to tilt and twist boundaries as discussed in [7]. In addition, the length scale of misoriented boundaries is observed to range from a few nanometers to hundreds of nanometers, and these boundaries are frequently observed to intersect.

Partially dissolved particles:

In all cases, the ratio of benzoquinone to total dissolved Co is 1:2, as predicted by eq. 1. Control experiments involving hydroquinone solutions without the presence of heterogenite particles showed no oxidation to occur. The mass balance between total dissolved Co and oxidized hydroquinone indicates that no sorption or precipitation of Co-bearing species occurs. TEM results confirm that no new phases are formed in partially dissolved samples.

Dissolution of these micron-sized plates occurs by two physically distinct mechanisms. The first involves dissolution over the entire surface of the plate (figure 4) and the second involves preferential dissolution along misoriented boundaries between primary heterogenite nanocrystals (figure 5). Considering misorientations between primary nanocrystals, we suggest that boundaries of misorientation represent a disproportionate amount of the total surface reactivity in comparison to the total physical surface area. HRTEM results clearly show slight crystallographic misorientations across channels along which material has been removed. However, the average thickness, as determined by taking the ratio of the zero-loss TEM image to the unfiltered TEM image (fig. 6), shows that the effective thickness of the undissolved material surrounding these channels is not significantly different from that of the undissolved particles. The thickness of preferentially dissolved regions

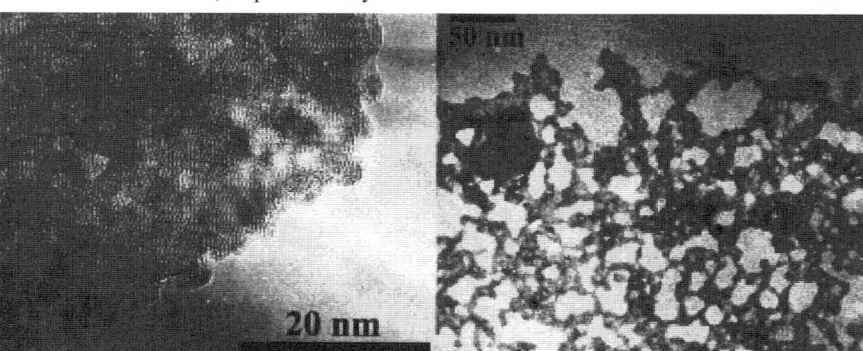

Figure 4: Zero-loss TEM image showing dissolution of CoOOH-2 particles. No pathways of preferential dissolution are observed over the entire surface of this plate.

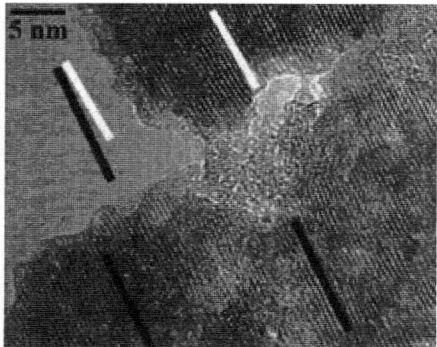

Figure 5: Zero-loss TEM images showing preferential dissolution along misoriented boundaries. Heavy lines serve to highlight the slight crystallographic misorientation across this channel along which material has been removed.

is commonly less than 1/2 the thickness of the surrounding material. Thus, the reactivity of misoriented boundaries is significantly higher than the reactivity in regions of perfectly oriented assembly.

This result is similar to observations of preferential dissolution along dislocations and planar defects as seen in natural samples of alkali feldspars [8]. These microstructural features appear to dramatically alter the relative reactivity of a material's surface. Finally, natural heterogenite from Goodsprings, Nevada, (NMNH # 94423-5) exhibits a microstructure similar to that of this synthetic heterogenite (fig. 7). These misoriented boundaries are expected to dominate the effective reactive surface area of this material and to strongly impact the fundamental mechanisms by which heterogenite dissolves in nature.

CONCLUSIONS

Dissolution occurs primarily along misoriented boundaries between the primary building blocks of heterogenite plates formed via oriented assembly. This work is a first step toward linking microstructural features and crystal growth mechanisms with specific chemical behavior by coupling atomic-scale observations with wet chemical methods.

ACKNOWLEDGMENTS

The authors wish to gratefully acknowledge funding from the National Science Foundation (EAR 9418090) and the U.S. Environmental Protection Agency National Center for Environmental Research and Quality Assurance (R82-6376). We also wish to thank the Smithsonian National Museum of Natural History for providing a sample of naturally occuring heterogenite from Goodsprings, Nevada (NMNH # 94423-5).

REFERENCES:

1. McArdell CS, Tien CF, and Stone AT, "Formation of Co(III)-iminodiacetate isomers on the surface of CoOOH (heterogenite)", in preparation.
2. Banfield JF and Hamers RJ, Geomicrobiology: Interactions between microbes and Minerals, **Vol. 35** of Reviews in Mineralogy, JF Banfield and KH Nealson, EDS (Mineralogical Society of America, Washington, DC), pp. 86-122 (1997).
3. Means JL, Crerar DA, Duguid JO, *Science*, **200**, 1477-1481 (1978); Brusseau ML and Zachara JM, *Environ. Sci. Technol.*, **27**, 1937-1939 (1993)
4. Stone AT and Ulrich HJ, *J. Col. Interf. Sci.*, **132**, 509-522 (1989).
5. For details regarding EFTEM techniques, see Reimer L, Energy-Filtering Transmission Electron Microscopy, **Vol. 71** of the Springer Series in Optical Sciences, L. Reimer, ED (Springer-Verlag, Berline, Heidelberg), pp. 347-400 (1995).
6. Delaplane RG, Ibers JA, Ferraro JR, and Rush JJ, *J. Chem. Phys.*, **50**, 1920-1927 (1969).
7. Penn RL and Banfield JF, *Science*, **281**, 969-971 (1999).
8. Hochella MF and Banfield JF, Chemical Weathering Rates of Silicate Minerals, **Vol. 31** of Reviews in Mineralogy, AF

White and SL Brantley, EDS (Mineralogical Society of America, Washington,
DC, 1997), pp. 353-406 (1995).

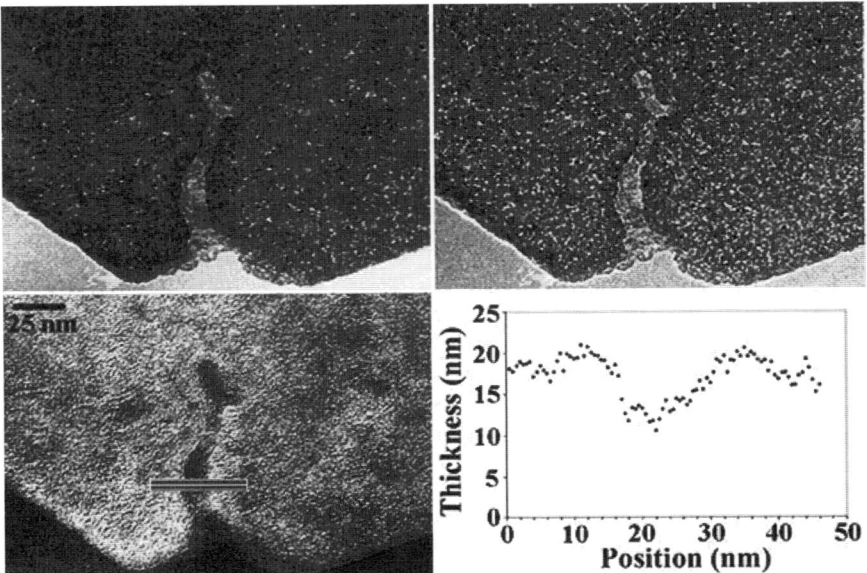

*Figure 6: Zero-loss (upper left) and unfiltered (upper right) TEM images of a plate after
equilibration with hydroquinone solution. The ratio of the zero-loss image to the
unfiltered image, shown in the lower left image, allows a measure (lower right) of the
relative thickness of material removed along these pathways.*

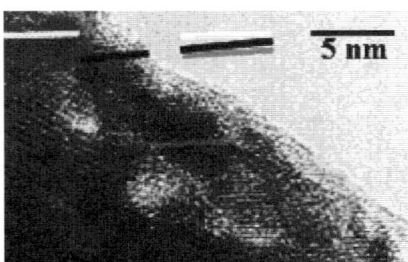

*Figure 7: Zero-loss TEM image of nat-
ural heterogenite from Goodsprings,
Nevada (Smithsonian NMNH # 94423-5).*

Poster Session:
Morphology and Dynamics
of Crystal Surfaces in
Complex Molecular Systems

Mat. Res. Soc. Symp. Vol. 620 © 2000 Materials Research Society

Formation of Cellular Crystalline Submicrostructure in the Butter with Additives

Rashevskaya T. A., Gulyi I. S.
Ukrainian State University Of Food Technologies,
68 Vladimirskaya Str., Kiev-33, 02033 Ukraine

Nishchenko M. M., Likhtorovich S. P.
G. V. Kurdyumov Institute for Metal Physics, N.A.S.U.,
36 Vernadsky Blvd., Kiev-142, 03680 Ukraine

Introduction

Ecological situation of the last decades has driven an interest to the production of nutrients with natural food supplements. We have developed a number of cream-butter sorts containing vegetable additives, which are recommended as the dietary and prophylactic food supplements [1]. These kinds of butter exhibit unusual plasticity and enhanced durability, which are attributed to the features of their microstructure formation.

The developed technology can also be applied to manufacture other fat-containing nutrients: spreads, margarines, pastes, etc. Thus, the research on the influence of vegetable additives on the formation of submicrostructure in the fat-containing products with complex multicomponent composition is important from both theoretical and practical viewpoints.

Earlier results on the microstructure of the butter were based on its microscopic studies in polarized light [2]. The observed refraction was accounted for the presence of 'the needle-like crystals in the boundary region of the butter's fat globules'. The electron microscopy has been applied to the studies of butter since 1960s [3, 4], when the ultrathin cuts of chemically fixed butter or the replicas of cleavage-surface were used. These techniques could not guarantee exact reproduction of the butter's microstructure and clear imaging of its elements. Since 1970s, the use of rapidly frozen and then cleaved samples has enabled one to obtain the images of real microstructure and rather sharp reproduction of its elements.

The surface of the fat globules and the interglobular regions were found to consist of small crystalline aggregates formed by monomolecular glyceride layers [5].

In present paper the influence of the red-beet cryopowder additive (obtained by means of sublimation drying) on the formation of the fat-phase submicrostructure in the butter has been studied using the electron microscopy.

Methods of investigation

The butter samples for the electron-microscope studies were prepared by means of so called cooling-and-freezing technique: they were rapidly cooled to the liquid-nitrogen temperature and then cleaved in a vacuum of 10^{-4} Pa. Such a method allowed us to fix the butter's structure existing at its conservation temperature. Thin platinum-carbon layer was deposited onto the cleavage surface and the replica thus obtained was subsequently studied in the electron microscope.

The following samples of butter with and without the red-beet additive were studied (the former served as the reference samples): (*i*) in the as-prepared state; (*ii*) after 6-months conservation at $-18°C$.

Results and discussion

The electron-microscope images of the microstructure in the butter without additives are shown in Fig. 1. Structure of the as-prepared reference sample is considered to be homogeneous [3], since it contains destroyed and partially fractured fat globules 1.4 – 3.5 μm in size. They are submerged into the amorphous-crystalline fat phase (Fig.1, a). The globules' surfaces are covered with 4 – 5 nm thick monomolecular crystalline layers closely adjoining each other. In the interglobular regions the variety of lamellar crystal aggregates are observed. They are 1.0 to 50 μm in size and posed at different angles to each other. Lamellar crystals consist of flat monomolecular glyceride layers.

In the course of conservation of the butter without additive, two concurrent processes occur: exfoliation of lamellar crystals to monomolecular layers and separation of crystalline layers into smaller crystals of (irregular) polyhedral shape. With increasing duration of conservation and decreasing temperature the processes of crystalline layers' separation and their further exfoliation to separate monomolecular glyceride layers are enhanced, as it is clearly seen in Fig. 1, b.

Microstructure of the as-prepared butter with the red-beet additive contains a lot of undestroyed fat-globules and is, therefore, characterized as 'granular' [3]. They are 1.4 to 4.2 nm in size (Fig. 2, a), that is larger than in reference samples. Small rounded structures looking like knobs and cavities (about 60 nm in size) are visible on the surface of fat globules as well as in interglobular regions.

Microstructure of the butter with red-beet additive after conservation also contains a lot of undestroyed fat-globules. On the surface of the globules' crystalline envelopes and in interglobular spaces the cellular structure has formed (Fig. 2, b).

Formation of the cells is related to the processes of crystallization of low-melting glycerides, recrystallization and phase transformations of previously solidified glycerides. With increasing duration of storage and decreasing temperature the cells' size increases and at the storage temperature of $-18°C$ their mean size is about 100 nm.

The most low-melting components of the milkfat and ultradisperse water nanoparticles are distributed over the cells' boundaries. Formation of cellular morphology is related, in the first place, to compositional diversity of triglycerides that results in the wide range of their melting points (from – 40 to +40°C). In the second place, the variety of admixtures dissolved in the butter's liquid phase with red-beet additive also plays a role. The liquid phase in the butter consists of the liquid fat with emulsified water. Components of the red-beet cryopowder, along with lactose, proteins and other organic compounds, are dissolved in the water phase. The butter also contains adsorbate (bound) water, which has been identified in [6].

The following mechanism is suggested to explain formation of the cellular structure. At the initial stage the knobs are formed. Upon their nucleation in the butter, crystallization centers consist of the most high-melting glycerides of the fat phase, because they are characterized by the strongest intermolecular bonds and experience the largest undercooling. In the process of crystallization the fat components are selected according to their undercooling degree and the strength of intermolecular bonds [7]. At the same time, composition of the milkfat liquid phase is enriched with components, which are pushed out of the growing crystals. They are pushed outwards by crystallization front due to diffusion and convection. This process is clearly reflected in Fig. 3 that shows magnified patch of the microstructure in the as-prepared butter, presented in Fig. 2, a.

Groups of the milkfat glycerides solidify gradually and in succession, as they approach their undercooling threshold. So they form crystalline layers around the nuclei which consist of the high-melting glycerides. Compounds dissolved in the liquid phase are accumulated at the crystallization front. In fact, they do not affect the kinetics of nucleation, but essentially influence further growth of the crystals, slowing down this process and changing crystals' morphology. The crystals have the shapes of octahedra, dodecahedra or icosahedra. Interestingly, the water nanoparticles are arranged at their vertices and edges. At crystals' vertices they are 12 to 16 nm in diameter, while at the edges they are somewhat smaller (8 to 12 nm). Therefore, formation of the water nanoparticles is an heterogeneous process. Their formation at the edges and the vertices of crystalline polyhedra leads to the decrease in the surface energy. Thermodynamically stable are spherical nanoparticles of water with critical radius, which are stabilized by compounds of the red-beet additive.

Therefore, polyhedral cells in the structure of butter form crystalline network filled with the most low-melting components of the milkfat and the nanoparticles of the water phase. This submicrostructure hampers formation of cracks and fractures in the butter and results in its increased plasticity and spreadability.

Conclusions

1. It is shown that in the course of storage of the butter with red-beet additive at -18°C the cellular structure is formed on the surface of globules and in the interglobular regions. The size of the cells is about 100 nm.
2. Central part of the cell consists of the most high-melting glycerides of the fat phase. Glycerides with lower melting points crystallize near the surface of the cell. The most low-melting components of the milkfat and the water nanoparticles are distributed in the interfaces between the cells.
3. Thermodynamically stable water nanoparticles are formed heterogeneously, at the vertices and edges of crystalline polyhedra, leading to the decrease of the surface energy.

REFERENCES

1. Rashevskaya T. A. *et al.*, The use of novel sorts of cream butter enriched with vegetable cryopowders for medical and prophylactic nutrition.- Kharchova Promyslovist', 1998, No. 43, p. 67 – 70 (in Russian).
2. Mulder H., Walstra P. The milk fat globule. - Emulsion Science as Applied to Milk Products and Comparable Foods. Commonwealth Agric. Bureau, U. K. Farmham Royal, England, 1974.
3. Knoop A., Knoop E., Wortmann A. - Elektronenmikroskopishe Untersuchungen uber die phisikalische Struktur der Butter und Margarine. XVIIth Int. Dairy Congr., Munchen, 1966.
4. Precht D., Bucheim K. Elektronenmikroskopishe Untersuchungen uber die phisikalische Struktur von Streichfetten. 1.Die Mikrostruktur der Fettkugelchen. Milchwissenschaft. – 1979. – H. 12. S. 745 – 749.
5. Precht D., Bucheim K. Elektronenmikroskopishe Untersuchungen uber die phisikalische Struktur von Streichfetten. 2. Die Mikrostruktur der Zwischenglobularen Fettphase in Butter.-Milchwissenschaft, 1980, H.7.S.393 – 398.
6. Rashevskaya T. A. *et al.*, The role of water phase in formation of microstructure of butter with red beet powder additive. Int. Symp. on Water Management in the Design and Distribution of Quality Foods, 30 May – 4 June 1998, Helsinki, Finland. Proc. of Poster Sessions, p. 202 – 203.
7. Chalmers B. Theory of Solidification. Moscow, Metallurgia, 1968 (Russian translation).

Fig. 1

Fig. 1. Microstructure of the butter without additives: (a) as-prepared state; (b) stored at -18°C for 6 months. Indicated are the fat globule (A) and the layered structure in the interglobular regions (B).

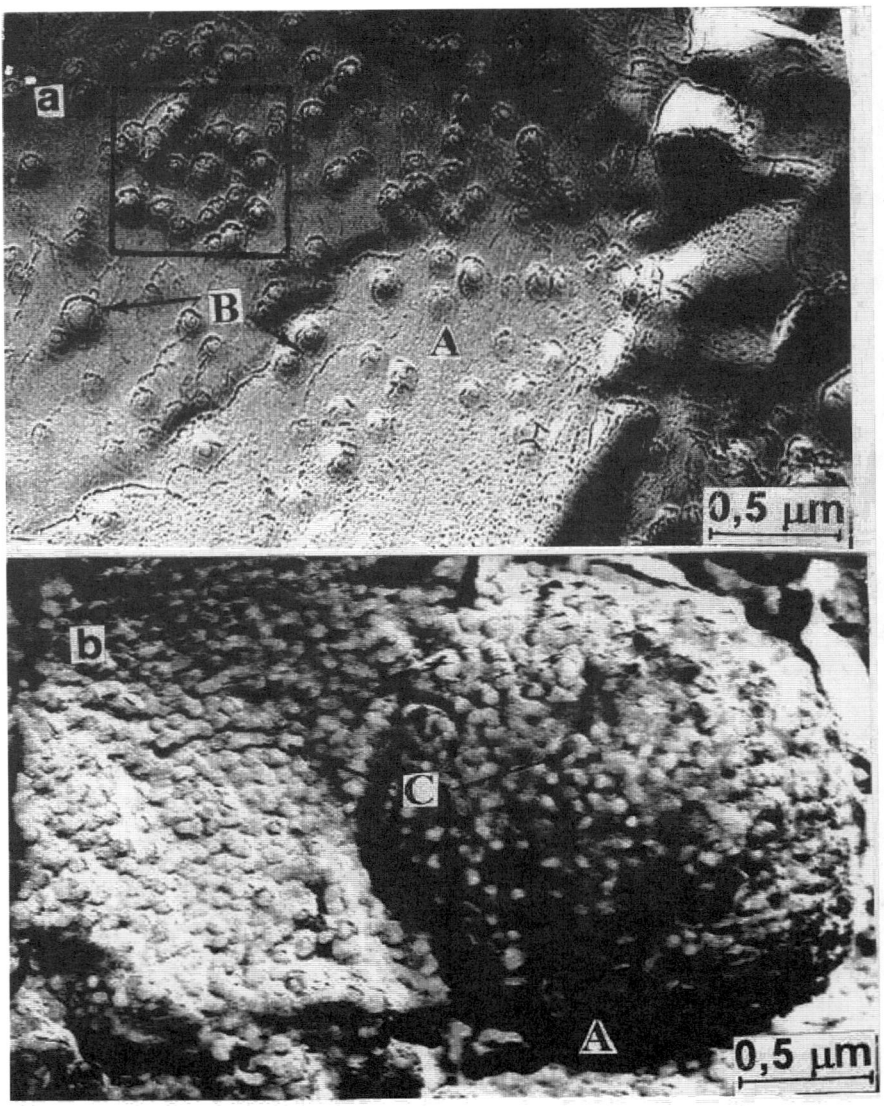

Fig. 2

Fig. 2. Microstructure in the butter with the red-beet cryopowder additive: (a) as-prepared state; (b) stored at -18°C for 6 months. Indicated are the fat globule (A), the knobs (B), and the cells (C).

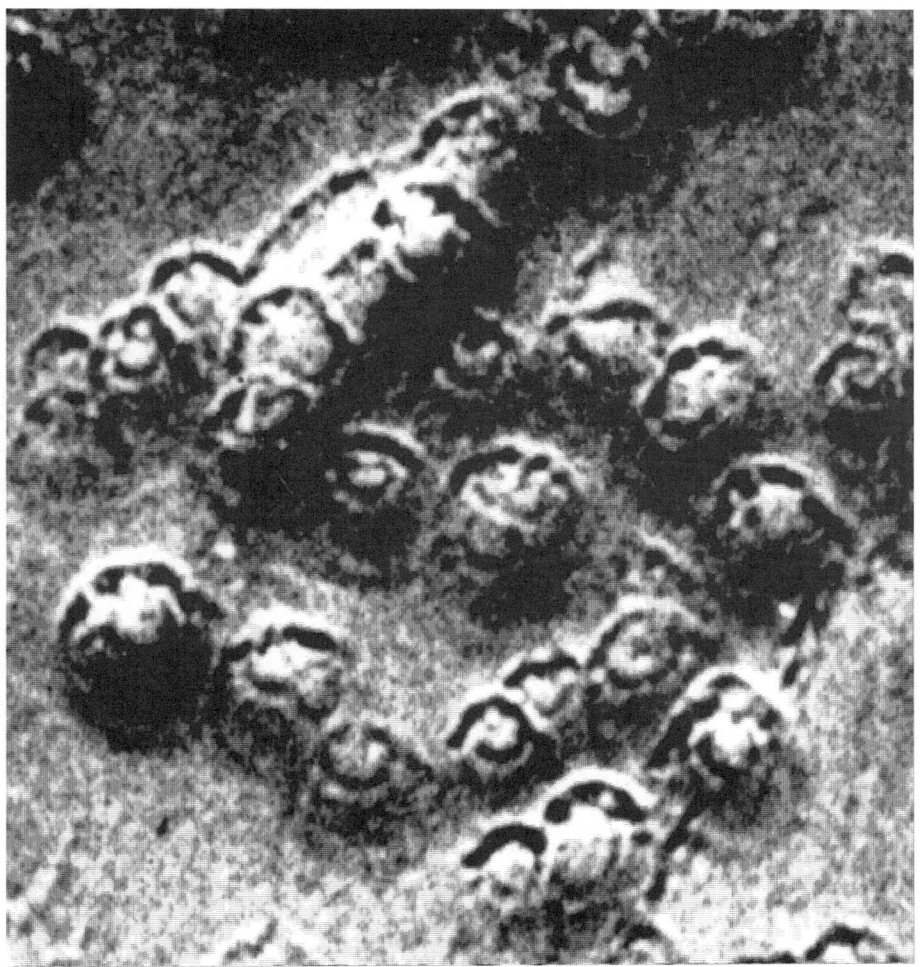

Fig. 3

Fig. 3. Magnified patch of the microstructure in as-prepared butter with red-beet additive (part of the image shown in Fig.2).

Mat. Res. Soc. Symp. Vol. 620 © 2000 Materials Research Society

Influence of external temperature field to period of eutectic pattern.

Gus'kov A.P.
Institute of Solid State Physics
Academy of Sciences of Russia,
Moscow District, Chernogolovka,
142432, Russia

INTRODUCTION

We used a mathematical model of crystal growth of [1] for the description of an eutectic pattern [2]. But the stationary problem gives physically unrealizable solutions (fig. 1) for values of system parameters corresponding to real experiments. Now known models [1,3-7] of crystal growth, used for a research of the interface stability of directed crystallization, not take into account of an external temperature field too. As examples of application of these models the regimes of crystal growth are usually used. The purpose of this work - to construct a model of a binary melt crystallization taking into account an external temperature field. Within the framework of this model we deduce analytical dependence of period of the eutectic structure on parameters of the system. We demonstrate, that for real parameters of the system, the parameters of the external temperature field weakly influence period of the eutectic pattern. This outcome is observed in experiments. We also explain the reason of joint emerging eutectic and dendrite of a structure for want of growth of eutectic crystals. We also explain the reason of joint both eutectic growth and dendrite growth under eutectic growth.

THEORY

Let $T(y,z,\tau)$ be the temperature normalized to the phase transition temperature T_{e0} and at initial impurity concentration C_0; $C(y,z,\tau)$ is the impurity concentration normalized to the initial one; y,z,τ are the dimensionless coordinates and the time: $y=\alpha y_r$, $z=\alpha z_r$, $\tau=\alpha^2 \chi_0 \tau_r$; D is the dimensionless factor of diffusion in a melt, $D = D_r/\chi_0$; $\chi = \chi_r/\chi_0$ is the factor of thermal diffusivity, ε is the heat of phase transition normalized to the specific heat capacity and the temperature of phase transition. $y_r, z_r, \tau_r, D_r, \chi_r, \varepsilon_r$ are dimensional quantities, $\chi_0=10^{-5}m^2s^{-1}$, $\alpha=10^2$ m^{-1}. We take into account the heat conduction in solid and liquid phases and diffusion of impurity in the liquid phase. To reduce the calculations in the equations we do not write down the coordinate x. The values relating to the solid phase are designated by stroke.

We describe the external temperature field $T_{ext}(z)$ by function of internal sources as exponential function

$$T_{ext} = \begin{cases} T'_{ext}(\infty) + \left(T_{ext}(\infty) + T_x - T'_{ext}(\infty)\right) \cdot \exp\left(\dfrac{\phi_0}{\left(T_{ext}(\infty) + T_x - T'_{ext}(\infty)\right)}z\right) \\[2em] T_{ext}(\infty) + T_x \cdot \exp\left(\dfrac{\phi_0}{T_x}z\right) \end{cases} \tag{1}$$

The parameters of function (1) satisfy to following conditions: At point at infinity the temperature values are $T'_{ext}(-\infty)$ and $T_{ext}(\infty)$; the sewing of function (1) is made at the point z = 0, at this point $T'_{ext}(0) = T_{ext}(0)$ and temperature gradient are equal to specific values ϕ_0. The constant T_x is searched from boundary conditions. We do not give here setting and solution of the stationary problem. The solution of one is similar to the solution of a problem of the heat conduction with external field, linear on the spatial coordinate, which explicitly is studied in [8].

For the problem, without external temperature field there is a restriction on magnitude of the temperature gradient on the interface. Let's set values of thermal parameters, which correspond to a solidification of a melt $Al_2O_3 + ZrO_2$: $\chi_r = 5 \cdot 10^{-6} m^2 s^{-1}$, $\chi / \chi' = 1.5$, $\varepsilon_r = 10^6 J \cdot kg^{-1}$, $D_r = 10^{-8} m^2 s^{-1}$. Value of heat transfer coefficient we take equal $\phi_r = 10^6 s^{-1}$. We take the values of experimental parameters $V_S = 10^{-5} m \cdot s^{-1}$, $T'_{ext}(-\infty) = 1900K$, $T_{ext}(\infty) = 2100K$, $T_{e0} = 2000K$ corresponding to the regime of crystal growth.

We set value of the temperature gradient ϕ_{0r} for the problem with the external temperature field. Let stationary solution of the problem has the temperature gradient on the interface in the solid phase. We compare this solution with a solution of the problem without the external temperature field with the same value of the temperature gradient on the interface in the solid phase. Let's consider dependence of stationary temperature close to the interface for two values of the gradient of the external field (fig.1).

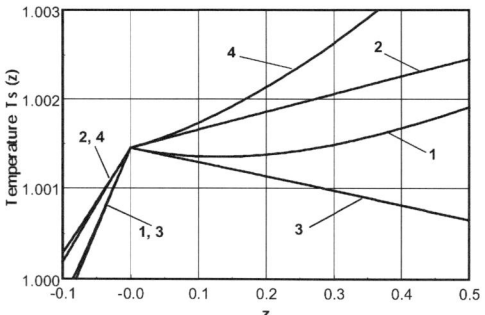

Figure 1. *Dependence stationary temperature on z. The problem with – 1, 2, and without - 3, 4 external temperature field. 1, 3 - $\phi_0 = 10^3 K\, m^{-1}$, 2, 4 - $\phi_0 = 2 \cdot 10^3\, K\, m^{-1}$.*

In experiments on crystal growth the temperature gradient of the external field close to the interface has usually values in the range $10^3 - 10^4 K\, m^{-1}$. The Figure 1 shows, what exactly in this range the solution of the stationary problem without the external temperature field loses physical sense - temperature in the liquid monotonically decreases. Such behavior formally is clear - the solution of the problem contains only one of exponential curve, which gives only monotone solutions. Therefore correctly formulated problem should take into account the external temperature field.

It is known [9], that in experiments the growth of the eutectic pattern is accompanied frequently by dendrite growth. Some articles explain growth of dendrites by small value of the temperature gradient before interface [8]. Our calculations explain simultaneous growth of dendrites and eutectic pattern. If the gradient of the external temperature field in the liquid before the interface is rather small, in crystal there are conditions for growth of dendrites. Now we shall show, that for given here parameters because of instability of the interface the eutectic pattern arises in the solid and we shall find period of this pattern.

Let's insert the external temperature field (1) into the model of [1]. The boundary value problem for small perturbations in curvilinear frame attached to the interface [1], has form

$$\chi' \frac{\partial^2 T'_m}{\partial z^2} + V_S \frac{\partial T'_m}{\partial z} + \left(\chi' K^2 - \omega - \phi \right) T'_m = \frac{\left(\chi' K^2 - \omega \right)}{\omega} V_m \frac{\partial T'_S}{\partial z}; \quad -\infty < z \leq 0; \quad (2)$$

$$\chi \frac{\partial^2 T_m}{\partial z^2} + V_S \frac{\partial T_m}{\partial z} + \left(\chi K^2 - \omega - \phi \right) T_m = \frac{\left(\chi K^2 - \omega \right)}{\omega} V_m \frac{\partial T'_S}{\partial z}; \quad 0 \leq z < \infty; \quad (3)$$

$$D \frac{\partial^2 C_m}{\partial z^2} + V_S \frac{\partial C_m}{\partial z} + \left(DK^2 - \omega \right) C_m = \frac{\left(DK^2 - \omega \right)}{\omega} V_m \frac{\partial C_S}{\partial z}; \quad 0 \leq z < \infty; \quad (4)$$

$$\chi' \frac{\partial T'_m}{\partial z} \bigg|_{z = 0 - 0} - \chi \frac{\partial T_m}{\partial z} \bigg|_{z = 0 + 0} = \varepsilon V_m \quad T'_m \bigg|_{z = 0 - 0} = T_m \bigg|_{z = 0 + 0} \quad T_m \bigg|_{z \to \infty} = 0; \quad (5)$$

$$D \frac{\partial C_m}{\partial z} \bigg|_{z = 0 + 0} = (1 - k) \left(V_S C_m + C_S V_m \right) \quad C_m \bigg|_{z \to \infty} = 0; \quad (6)$$

$$V_m = \theta T_{m0} + \gamma C_{m0} ; \quad \theta = \frac{\partial V}{\partial T} \bigg|_{T = T_S(0)} ; \quad \gamma = \frac{\partial V}{\partial C} \bigg|_{C = C_S(0)} ; \quad (7)$$

where k - segregation coefficient. We assume, that the solutions of the problem have form [1]

$$T' = T'_S(z) + T'_m(z) \exp(\omega \tau + Ky); T = T_S(z) + T_m(z) \exp(\omega \tau + Ky);$$
$$C = C_S(z) + C_m(z) \exp(\omega \tau + Ky)$$
$$\omega = \omega_1 + i\omega_2 ; \quad K = K_1 + iK_2 ;$$
$$T'_m(z) << T'_S(z); \quad T_m(z) << T_S(z); \quad C_m(z) << C_S(z);$$

here $T'_S(z)$, $T_S(z)$, $C_S(z)$ - solutions of the stationary problem.

RESULTS

To obtain the dispersion equation of the system we find the solution of the problem (2) - (7). On the interface this solution gives a linear set of equations concerning factors $T_{m0} = T_m(0)$ и $C_{m0} = C_m(0)$

$$\left(S'_T - S_T \right) T_{m0} - \eta V_{m0} = 0 \quad (8)$$

$$[S + 2(1-k)]C_{m0} - (k-1)\xi V_{m0} = 0 \quad (9)$$

here S'_T, S_T, S – the radicals of characteristic equations of the system (2) – (4), η and ξ depend on parameters of a system, and $\xi \neq 0$ at $k = 1$. Substituting expansion (7) in (8), (9) we obtain the dispersion equation as

$$\left(S'_T - S_T \right) - \eta\theta\left[S + 2(1-k) + (1-k)\xi\gamma\right] + (1-k)\xi\eta\gamma\theta = 0 \quad (10)$$

More in detail scheme of calculation of the dispersion equation of the system is resulted in [10]. Let's remark, that if the velocity of the interface does not depend on temperature and concentration, then $\theta = 0$, $\gamma = 0$ and in this case the problems of the diffusion and heat conductions are independent. In this case the equation (10) is degenerated into product of the dispersion equations of problems of heat conduction and diffusion. If the component concentration is close to eutectic, then $k \rightarrow 1$. If $1 - k$ rather small value, the equation (10) has form

$$\left(S'_T - S_T \right) - \eta\theta\left|S\right. = 0 \quad (11)$$

This equation decompose into a system of two real equations, appropriate to imaginary and real parts of the equation (11)

$$\begin{cases} \mathrm{Re}((S'_T - S_T) - \eta\theta)\mathrm{Re}\,S - \mathrm{Im}[(S'_T - S_T) - \eta\theta]\mathrm{Im}\,S = 0 \\ \mathrm{Re}((S'_T - S_T) - \eta\theta)\mathrm{Im}\,S - \mathrm{Im}[(S'_T - S_T) - \eta\theta]\mathrm{Re}\,S = 0 \end{cases} \quad (12)$$

imaginary part of S has form

$$\mathrm{Im}(S) = -0.5\sqrt{2\sqrt{(1+Y+\delta)^2 + \Omega^2} - 2(1+Y+\delta)} \quad (13)$$

The results of numerical calculations will be necessary for further simplifications. For this calculations we give values of segregation coefficient $k = 1.03$ and slope of liquidus $m = -0.05$ same, as in [2]. The value k in [2] simulated the phenomenon of lamination of the melt on the interface. Temperature gradient of the external field let's assume equal $\phi_{0r} = 10^4 K\,m^{-1}$. We use a model of growth with by means of screw dislocations for numerical calculation, with kinetic factor $h = 2.2 \cdot 10^{11}$. We write down the solution as the projection of the trajectory $\delta\,(\Omega,\,Y)$ on plane (δ, Ω) (curve 1), and (Ω, Y) (curve 2) (fig. 2)., where

$$\delta = \frac{4D\omega_1}{V_S^2}; \quad \Omega = \frac{4D\omega_2}{V_S^2}; \quad Y = \frac{4D^2 K_2^2}{V_S^2},$$

are increment of growth rate, frequency of temporary pulsations and frequency of spatial distortions.

These dependence show, that for specific parameters of the system, the interface has the spatial mode with frequency $log(Y) \approx 8.5$, as well as in the problem disregarding of external temperature field. According to growth rate hypothesis the system does not contain temporary pulsations, since the trajectory $\delta(\Omega)$ monotonically will increase when of $\Omega \to 0$. Therefore at the point of the maximum of the growth increment $\Omega=0$ and the expression (13) become

$$\text{Im}(S) = -0.5\sqrt{2\sqrt{(1+Y+\delta)^2} - 2(1+Y+\delta)} \equiv 0$$

It is easy to show that the expressions $\text{Im}\,S'_T$ and $\text{Im}\,S_T$ also will be converted to zero. Therefore in the system (12) at the point of the maximum of the growth increment there is only one equation, from which is obtained

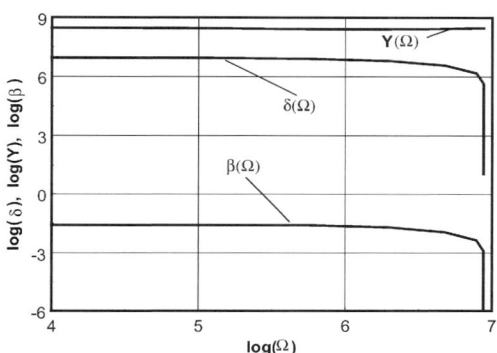

Figure 2. *Projection of trajectory $\delta(\Omega, Y)$ on planes $(\delta, \Omega) - 1$ and $(\Omega, Y) - 2$. Function $\beta(\Omega) - 3$.*

$$\text{Re}(S'_T) - \text{Re}(S_T) - \theta\,\text{Re}(\eta) = 0$$

Let's consider the expression $\text{Re}(S_T)$.

$$\text{Re}S_T = -1 + 0.5\sqrt{2\sqrt{\left(1+\frac{4\chi\phi}{V_S^2} + \frac{\chi^2 Y}{D^2} + \frac{\chi\delta}{D}\right)^2} + 2\left(1+\frac{4\chi\phi}{V_S^2} + \frac{\chi^2 Y}{D^2} + \frac{\chi\delta}{D}\right)}$$

We substitute here the parameters values and have obtained

$$1 << \frac{4\chi\phi}{V_S^2} << \frac{\chi\delta}{D} << \frac{\chi^2 Y}{D^2}$$

The similar inequalities are obtained and for expression $\text{Re}(S'_T)$. We can write down expression $\text{Re}(\eta)$ as

$$\text{Re}(\eta) = \frac{2\varepsilon}{V_S}(1+\beta)$$

Values of the parameter β along the trajectory $\delta(\Omega, Y)$ is shown on figure 2. From this figure follows, that $\beta << 1$. Neglecting in $\text{Re}(S_T)$, $\text{Re}(S'_T)$ and β by smalls and substituting them in (14) we obtain the equation

$$\left| \chi_1 + \chi \right| \sqrt{Y} - \frac{2\varepsilon D \theta}{V_s} = 0$$

The solution of this equation gives period of spatial distortions λ

$$\lambda = \frac{2\pi \left| \chi_1 + \chi \right|}{\varepsilon \theta}$$

The obtained expression coincides with the solution of the problem disregarding of external temperature field. For the model of growth by means of screw dislocations it gives dependence $\lambda(V_S)$ conterminous with experimental [2]. The deduction of this expression explains one more experimental result - temperature field practically does not influence period of the eutectic pattern for considered here system parameters.

CONCLUSIONS

1. The offered model of the crystallization explains emerging dendrites during the eutectic solidification.

2. From this model we have obtained the analytical dependence of the eutectic pattern period on the interface velocity, which coincides with experimental.

3. The model also explains, why in experiments the influence of the gradient of the external temperature field to period of the eutectic pattern is not observed.

4. Also there are all bases to assume, that the magnitude of the temperature gradient of the external field influences intensity of dendrite growth. If the temperature gradient of the external temperature field is increased, the intensity of dendrite growth should decrease, because the temperature gradient before the interface is increased.

Author thanks senior scientist. M. Starostin for useful discussions of calculations outcomes.

REFERENCE

1.   A.P.Gus'kov. Physics - Doklady v.**366** №4. (1999) p.468-471.
2.   A.Gus'kov. Electronic Journal "Investigated in Russia", 28, 1999,
      http://zhurnal.ape.relarn.ru/articles/1999/028.pdf
3.   Mullins,W.W., Sekerka,R.F.,// Journ. Appl. Phys., 1964, T.35, C.444.
4.   M.L. Frankel // Physica D, 1987, V. 27, P. 260-266.
5.   S.C.Hardy, G.B.McFadden, S.R.Coriell, P.W.Voorhees, R.F.Sekerka // Journal of Crystal Growth, 1991, V. 114, P. 467-478.
6.   David C, Sarocka. Andrew J. Bernoff, // Physica D, 1995, V. 85 P. 348-374.
7.   A.Gus'kov, Izvestija Akademii Nauk, Physics, v.63, №9, 1999, pp.1772 – 1782, (Russia).
8.   N.A.Avdonin, Crystal growth, Erevan, vol.XI, 1975, 268-272.
9.   R.Elliot, Eutectic Solidification Processing. Butterworths and Co Publishers Ltd., 1983.
10.   A.P.Gus'kov. Physics - Doklady v.**349** №4. (1996) p.468-471.

Macromolecules

Mat. Res. Soc. Symp. Vol. 620 © 2000 Materials Research Society

Growth kinetics and diffraction properties of STMV crystals

Yu. G. Kuznetsov, A. J. Malkin and A. McPherson
University of California, Irvine
Department of Molecular Biology and Biochemistry
Irvine, CA 92697-3900

ABSTRACT

Two crystal forms, orthorhombic and cubic, of satellite tobacco mosaic virus have been investigated. Atomic force microscopy showed that the orthorhombic crystals were characterized by a high density of point defects, while the cubic crystals were practically defect-free. Nonetheless, orthorhombic crystals diffract to a high resolution of 1.8 Å while the cubic crystals diffract to only about 4 to 6 Å resolution. Differences in the properties of viruses incorporated into the two crystal structures were demonstrated by growth kinetics studies. It appears that physical and chemical treatments applied to protein and virus solutions during their extraction and purification introduce a variety of specific structural changes and that these alterations may then affect the diffraction properties of resultant macromolecular crystals.

INTRODUCTION

The influence of crystal defects, such as point and linear defects, stacking faults, and grain boundaries present in X-ray diffraction data from inorganic crystals have been thoroughly studied. Results from these studies have been applied to macromolecular crystals to explain their marginal diffraction properties, but in most cases the arguments are not convincing. The sizes of the crystals and the estimated sizes of their blocks are, in general, large enough to yield X-ray data of high resolution. Furthermore, from AFM studies it is known that many macromolecular crystals are dislocation-free or have no more than a few dislocations. To improve macromolecular crystal quality, extensive purification of protein solutions (filtration, recrystallization), application of more precise growth techniques such as temperature control, and even microgravity conditions have been used [1-3]. Such efforts have improved matters in some cases, but not for all macromolecular crystals studied. In many cases there were no improvements at all.

Polypeptide chains making up a protein are twisted or folded to form a macromolecule with a specific three-dimensional conformation. Globular proteins composed of two or more polypeptide chains have four levels of organization referred to as primary, secondary, tertiary and quaternary structure. All levels use the same kind of bonds, which include hydrogen bonds, ionic, and covalent bonds, and hydrophobic interactions. Biological function, which is determined by the three-dimensional structure, can be altered either by replacement of an amino acid in the polypeptide chain sequence with another (mutation) or by altering the bonding pattern. Changes in shape and biological activity are also frequently observed when a protein is heated or treated with any number of chemicals.

The same kinds of interactions that are used to construct protein structure are also used between protein molecules to build crystal structure. Generally speaking, any physical or chemical treatment of a protein solution could affect protein structure as well as crystal structure.

It is known [1] that aberrant molecules may incorporate into protein crystals as an impurity and affect the crystal structure by promoting distortion of the crystal lattice.

Use of offensive chemicals, pH or higher temperatures during extraction of protein molecules and during crystallization can also disturb the structures of protein molecules and, as a consequence, their crystal structures. Deviation from the regular protein structures may be small but varied and randomly distributed among the virus or protein molecules. This in turn may affect the diffraction properties of the crystal.

In the present study, the crystal structure, growth kinetics and the influence of precipitant concentration on the defect structures of two satellite tobacco mosaic virus (STMV) crystal forms have been investigated by atomic force microscopy. These two crystal forms – orthorhombic and cubic, diffract very differently.

MATERIALS AND METHODS

Satellite tobacco mosaic virus has sixty identical subunits comprising its icosahedral capsid and a virion diameter of 17 nm [5,6]. The unit cell parameters of the orthorhombic crystal form, BCC, are a = 174.3Å, b = 191.8 Å, c = 202.5 Å. The unit cell parameters of cubic crystal form, FCC, are a = b =c = 257 Å. The orthorhombic crystals produce X-ray diffraction data to at least 1.8 Å resolution [7], while the cubic crystals barely diffract to 6 Å [not published]. For crystallization of the two crystal forms the same ammonium sulfate precipitant at the same concentration and pH = 7 was used. Usually freshly prepared virus solutions produce orthorhombic crystals. After prolonged storage however, cubic crystals begin to appear among the orthorhombic crystals until, ultimately, only cubic crystals grow from solution.

Crystal surface observations and measurements of growth step velocities were made with an atomic force microscope, tapping mode, Nanoscope III, (Digital Instruments, Santa Barbara, CA). To initiate experiments, crystal seeds were mounted on glass substrates under carbon fiber holders and placed in fluid cells. A fluid cell was then filled with solutions of various virus and salt concentrations.

EXPERIMENTAL RESULTS

Visualization by AFM consistently showed a high density of point defects in all orthorhombic STMV crystals, as in Figure1a. These crystals grow by two-dimensional nucleation and each successive layer, which spreads over a crystal face, has to penetrate between point defects. The crystals do not dissolve in water at pH = 7 which means that virus particles that incorporate into the crystal form bonds with total energy much more than kT. The growth kinetics of orthorhombic STMV crystals have been studied at low 0.015 – 0.25M, and high 2.5 – 4M ammonium sulfate concentrations. Due to the presence of the high defect density, growth step velocity fluctuated a great deal. As a consequence, growth step velocity was based on the period where the growth step had already passed one stopper but had not yet encountered another, i.e., during movement through small valleys between defects. Because of small valley size, and the probable presence of fluctuations in growth step movement, velocity values were scattered more than 100% under constant growth conditions.

In the ammonium sulfate concentration range of 0.06 – 0.25M and virus concentration of 0.1 – 0.7 mg/ml, the growth step velocity fluctuated around 8×10^{-10} m/sec. At 0.015M salt concentration and 0.7 mg/ml virus concentration, growth proceeded very slowly, particle by particle. To be certain that crystals do not grow from salt free virus solution alone, a crystal was placed in a solution with a high virus concentration of 8 mg/ml. No growth was observed.

At high salt concentration, 4M, and a virus concentration of 0.08 mg/ml many clusters of 2 – 4 virus particles were observed on the crystal surface but there was no growth. During scans some clusters detached from the crystal surface. This suggested either weak binding of the virus particles to the crystal surface or their failure to resist AFM tip pressure. Two-dimensional nuclei with 8 or more virus particles, and single virus particles attached to step edges for long periods of time were observed without change at lower salt concentrations. Thus detachment of virus particles is a consequence of weak binding between particles. It is useful to note, that such observations also demonstrate that AFM will allow imaging of small size clusters if they are present on crystal surfaces.

At 3M concentration of salt and 0.16 mg/ml concentration of virus particles, very seldom was incorporation of particles into growth step edges seen, and no two-dimensional nucleation was observed with an increase of virus concentration to 0.32 mg/ml, many two-dimensional islands, nucleated during first contact of the crystal with the freshly prepared solution. There was, however, no growth. Probably, an increase of virus density in solution produced multiple nucleation on the crystal surface as well as in the bulk solution. At 2.5M salt and 0.2 mg/ml virus, regular movement of growth steps as well as two-dimensional nucleation were observed. Growth of orthorhombic crystals has not been observed in the presence of solution, which produce cubic crystals.

Cubic crystals unlike orthorhombic crystals have a low density of point defects. Defects, which are observed, are formed by sedimentation of aberrant virus particles on the crystal surface, as seen in Figure 1b. These virus particles are larger in size than others. They adhere to the crystal surface, and after incorporation into the crystal, they remain surrounded by vacancies. Usually they remain for a long time on the crystal surface, even during scanning by AFM, completely surrounded by spreading growth layers. Similar kinds of particles were observed on surfaces of orthorhombic crystals.

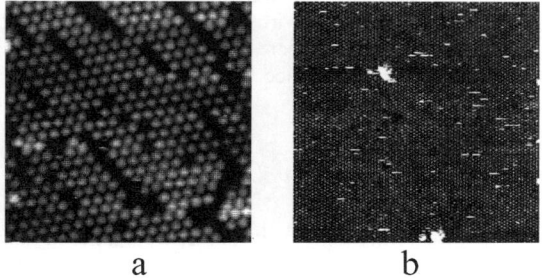

a   b

Figure 1. Surface structure of orthorhombic (a) and cubic (b) STMV crystals. The individual spherical virus particles, of 17 nm diameter, making up the crystals are evident in both images. Scan sizes are 400 x 400 nm² (a) and 1.5 x 1.5 μm² (b)

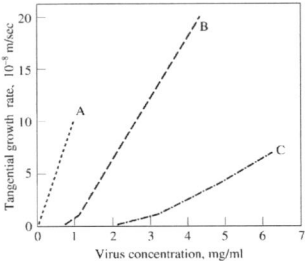

Figure 2. *Dependence of the tangential growth rate on virus concentration for cubic STMV crystals. Concentration of ammonium sulfate in a 1M, b- 0.4M and c- 0.29M*

Cubic crystals dissolve easily in water. At low salt concentrations solutions have equilibrium virus concentration, Figure 2. The higher the salt concentration the lower is the equilibrium concentration of viruses. Close to equilibrium it is difficult to determine with certainty if there is detachment of particles from growth step edges. At low salt concentrations, Figure 2, curve c, the crystal surface becomes sensitive to interaction with the AFM tip and a small increase in cantilever force leads to extraction of virus particles from step edges. Imaging of the growth step edge at virus concentrations only slightly higher than the equilibrium concentration, however, showed incorporation of particles into the step edge. On the other hand, there were no difficulties in imaging growth steps at high salt concentrations. The straight lines of steps with 1 and 4 particles attached to the edge were imaged for 20 minutes without changes.

A similar increase of crystal sensitivity to interaction with the AFM tip at low salt concentration was also observed for orthorhombic crystals. This implies that with an increase of salt concentration, the difference in chemical potentials for particles in the crystal and in solution increases. i. e. the strength of bonds between particles in the crystal increases. The same pattern was observed with regard to virus concentration – the lower the virus concentration the greater the crystal sensitivity. The rate of two-dimensional nucleation increases with virus concentration and the higher the salt concentration the stronger is the dependence.

DISCUSSION

Models of interaction

Despite a much higher density of point defects, orthorhombic crystals diffract far better than do cubic crystals. Virus particles in the orthorhombic crystals have four contacts with neighbors within the crystal surface (110) and two contacts with particles from the underlying crystal layer. In cubic crystals, a virion particle has six contacts within the crystal surface (111) and three contacts with particles from the underlying layer. The direction of these contacts for both crystal forms is reflected in the shapes of two-dimensional islands and growth step edges, Figure 3. It would appear that virus particles are better situated to maintain strong spatial order in the cubic crystal lattice. Cubic crystals, however, also exhibit less mechanical strength than orthorhombic crystals. Cubic crystals crack easily during mother liquor replacement with another at slightly lower salt concentration.

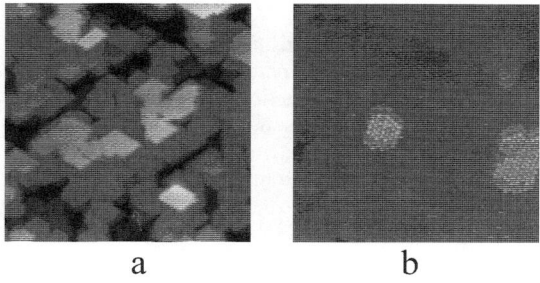

a   b

Figure 3. *Shapes of growth steps and two-dimensional islands reflect the directions of the strongest bonds between virus particles within orthorhombic (a) and cubic (b) crystal surfaces. Scan sizes are 1.1 x 1.1 μm² (a) and 880 x 880 nm² (b)*

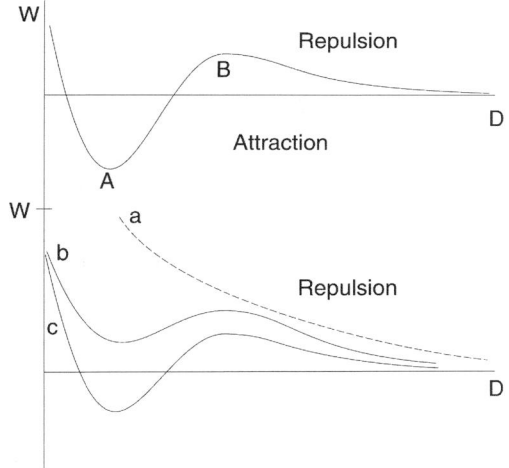

Figure 4. *Models of interactions between virus particles in solutions, which produce orthorhombic (a) and cubic (b) crystals*

The behavior of virus solutions can be described by DLVO theory [8]. That theory combines van der Waals and double-layer forces mutually acting. Depending on the properties of the particles and the solution in which they are dissolved, dependence of the potential energy of interaction between two particles in solution can vary. For solutions which produce orthorhombic crystals that do not dissolve in water we can apply the following model, Figure 4 a. In close contact, primary minimum A virus particles have strong van der Waals attraction to each other. For virus particles dissolved in water, the electrostatic repulsion barrier B prevents them from sticking to one another. The depth of the well in the primary minimum is so great that crystals do not dissolve in water and, clearly, thermal energy is insufficient for detachment of particles from the crystal. With an increase in distance between particles from the contact position, electrostatic repulsion prevails over van der Waals attraction and a barrier appears which prevents particles from sticking to each other in aqueous solution. Thus the virus solution is kinetically stable. The repulsion barrier does not allow the system to reach the lower energy level. The orthorhombic crystal neither dissolved nor grew in a solution having a virus concentration of 8 mg/ml.

To decrease the height of the barrier, salt ions are introduced into the virus solution. Higher salt concentration promotes contact as evidenced by a higher growth rate of the crystal at the same virus concentration.

It is important to note that protein molecules and virus particles do not have uniform shapes and electric charge distributions. Thus, when we consider interactions between particles in positions corresponding to those within the crystal lattice, any deviation from exact crystallographic position will decrease the energy of interaction between particles in the crystal. For example, at 4M salt concentration (58% saturated solution) the electrostatic repulsion is so strongly screened by salt ions that virus particles adhere to each other in random orientations. These small clusters, which were formed on or close to the crystal surface, were imaged by AFM, but destroyed during imaging because bonds between particles were as weak as bonds between particles and crystal. Salt ions associating with protein molecules produce not only a screening effect; their interaction is more complicated. Orthorhombic crystals in solution with concentrations of virus 0.2 mg/ml, ammonium sulfate 0.125M and sodium chloride 5% slowly dissolve.

For a virus solution that produces cubic crystals there is another model of interaction between particles, Figure 4b. In water, virus particles electrostatically repel each other, curve a. At high virus concentrations, when particles are very close to each other, there is probably a primary minimum on the diagram, but we did not concentrate the virus solution so extensively. At a salt concentration of 0.25M, however, it is possible to observe the primary minimum, the crystal phase, and repulsion barrier, Figure 4, curve b. At low concentrations, virus particles repel one another as in the previous model. When the potential energies of particles in the crystal and in solution become equal, we achieve saturation. It is interesting that interactions between particles at low salt concentrations remain repulsive, even in the crystal. Increase of virus concentration, movement along the distance axis from the right to the left, leads to a point where potential energies of virus particles in the crystal and in solution become equal – the equilibrium concentration. A further increase of virus concentration produces crystal growth. Increase of salt concentration reproduces the previous model (Figure 2, curve A). Now one can explain why cubic crystals so easily crack during replacement of one solution with another of slightly lower salt concentration. Of course, interaction between a particle and a crystal is not the same as between two particles. We extended the model of the particle - particle interaction to explain the particle – crystal interaction in the interest of simplicity.

CRYSTAL DEFECTS

An interesting interaction of cubic crystals with some unknown linear polymer also present in solution was observed. These polymers could be derived from microbes, which contaminated the virus solution at some point, and may be actin microfilaments. These linear polymers sediment on the (111) cubic crystal surface and arrange themselves in the valleys between virus particle lattice rows. Thus they are always observed to assume one of three symmetry related orientations on the crystal surface. At high salt concentrations a growth step encountering the polymer forces it upward by growing underneath it. Thus the chains are not incorporated into the crystal under high salt conditions. At low salt concentrations, however, the interaction between the polymers and the crystal surface changes. Growth steps do not lift the polymers but surround them. Each chain thereby produces a permanent defect on the crystal surface, Figure 5. Many crystal layers are then required to heal the scar. Thus the chains produce large inclusions of solution in a crystal. The polymers have never have been observed on orthorhombic crystal surfaces. Possibly they are repelled from that crystal surface.

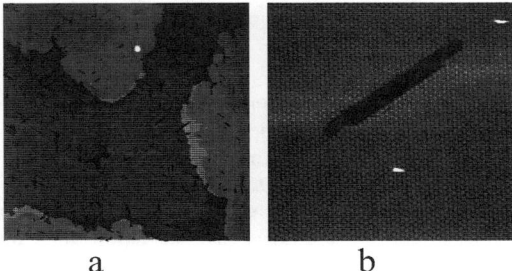

a   b

Figure 5. *Crystal defects formed on the surface of cubic STMV crystals due to incorporation of protein polymers. Scan sizes are 25 x 25 μm^2 (a) and 1 x 1 μm^2 (b)*

The formation of extensive inclusions in cubic crystals is further evidence of weak bonds between particles in comparison with orthorhombic crystals. There, each new crystal layer is not an exact replica in terms of defect pattern of the underlying crystal layer. Some defects propagate from the underlying layer, some are covered and some new defects appear.

Some cubic crystals contain defects like those seen in Figure 6. Probably these appeared as a result of stress relaxation. They are screw dislocations around a core about which growth steps rotate. For an inorganic crystal dislocations of such large core size would be highly unusual, even impossible. The screw dislocation, shown in Figure 6 a, has the Burgers' vector equal to five unit cell parameters. The dislocation was observed over a long time period without demonstrating any attempt to split. Splitting of single dislocations should, in principle, decrease the energy of the crystal, but for macromolecular crystals exhibiting low energies of interaction between particles, such dislocations are clearly tolerable. Screw dislocations like those observed in cubic STMV crystals have also been seen on thaumatin (Figure 6 c), canavalin [9] and other protein crystals. Screw dislocations having nine unit cell parameters of Burgers' vector have been recorded from thaumatin crystals. Orientations of dislocation cores are, of course, always correlated with crystal structure. They should have positions in the crystal plane perpendicular to the directions of the weakest bonds between molecules in the crystal.

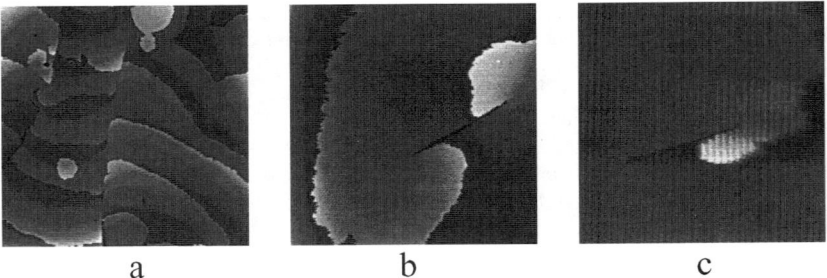

a   b   c

Figure 6. *Screw dislocations in STMV cubic crystals showing enlarged cores with Burgers' vector 5 unit cell parameters (a) and 2 (b), and in a thaumatin protein crystal of 1 (c). Scan sizes are 40 x 40 µm² (a), 6.4 x 6.4 µm² (b) and 600 x 600 nm² (c)*

The sensitivity of the cubic crystal lattice to lattice distortion is demonstrated in Figure 7. Growth steps cannot propagate between dislocations the distance between which is 1µm. Figure 7a. The growth step will pause until another growth step merges with it, Figure 7 b – d.

At molecular resolution one can observe many imperfections in the crystal lattice at the junction where two steps converge. In some instances, growth steps surmount obstacles by formation protrusions, Figure 8., which is similar to one-dimensional nucleation where distortion is minimal on the growth step edge.

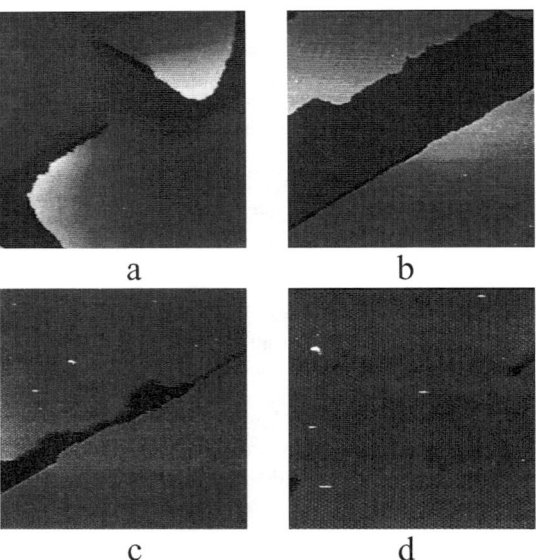

a   b

c   d

Figure 7. *Growth layers 1µm distant from one another on an STMV cubic crystal surface cannot penetrate between defects (a). The growth step pauses until another approaches and merges with it (b) and (c). No visible distortion of the crystal lattice between defects is observed (d). Scan sizes are 5 x 5 µm² (a) and (b), 2 x 2 µm² (c) and 1.3 x 1.3 µm² (d)*

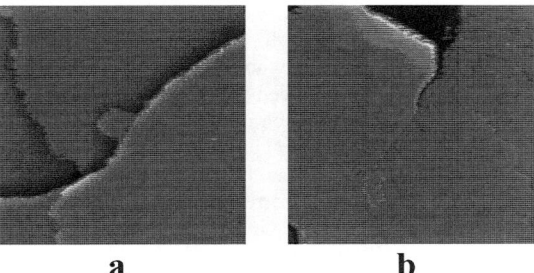

a   **b**

Figure 8. Growth layers penetrate between defects on a cubic STMV crystal surface by formation of protrusions on the step edge. Scan sizes are 2.3 x 2.3 μm² (a) and 5 x 5 μm² (b)

INFLUENCE ON CRYSTAL DIFFRACTION PROPERTIES

AFM images of cubic crystals, Figure 9 a, reveal virus particles having slightly different contrast, a reflection of virus height. Even more clearly, this contrast difference can be seen in images of bromegrass mosaic virus (BMV) crystals, Figure 9 b. Experiments with imaging of virus particles on cleaved mica surfaces also showed there to be differences in the sizes of virus particles. To increase attachment of virus particles to the mica surface, droplets of very dilute virus solution were dried on fresh cleaved mica. Imaging of dried virus was carried out under propanol. It was noted that the virus particles were sensitive to the drying process. They were deformed in the direction perpendicular to the mica surface, probably by capillary forces compressing them during drying.

The size of STMV particles from solutions producing orthorhombic crystals declined from 17 nm to 15 – 16 nm. For solutions producing cubic crystals the change was greater – the average size was 13 nm. During preparation and storage of virus we can expect that some particles will be damaged in various ways. Indeed, as noted above we frequently observed sedimentation of aberrant virus particles on the crystal surface, which failed to be incorporated. But we might expect that only virions with perfectly identical properties would be involved in the crystallization.

A few cubic crystals were separated from the solution and repeatedly washed in salt solution. Following that, the crystals were transferred into a droplet of water and dissolved. This droplet was then dried on the mica surface, Figure 9 c. One can see, however, the extensive differences in the sizes of the virus particles, all of which were at one time present in the crystal. Our next aim is to examine the sizes of virus particles coming from the crystal without drying. Drying may simply have demonstrated that particles in the crystal had different mechanical properties. Virus particles from solutions, which produce orthorhombic crystals, are more stable and uniform, Figure 9 d. This may offer an explanation as to why solutions, which produce orthorhombic crystals, following long storage, begin to produce cubic crystals. Properties of virus particles change during storage.

In a subsequent experiment, a freshly prepared virus solution, which produced orthorhombic crystals, was exposed to high pH = 9.5. After a few hours the solution was dialyzed for 24 hours against water. Crystallization yielded cubic crystals. The new cubic crystals diffracted as poorly as previous ones. This suggests that treatment with high pH produces changes in the virus particles. These changes could be structural or electrostatic or both. Changes which occurred to individual virions during storage or during treatment with pH may not be exactly the same for every particle. In addition to a fundamental change which produced transition from one crystal form to another, there are likely to be other unique changes for each particle. It is the accumulation of these sets of unique changes that conspire to affect crystal diffraction properties.

The good diffraction properties of the orthorhombic crystals can be explained by strong interactions between virus particles where any discrepancy with correct position of the particle in the crystal structure is resolved by defect formation. In addition, virus particles making up the orthorhombic crystals are less varied and have more uniform properties.

It seems quite probable that many chemical treatments during purification and separation of biological molecules or assemblies may create structural changes. In the cases of proteins and viruses these changes likely affect the diffraction properties of crystals grown from these particles.

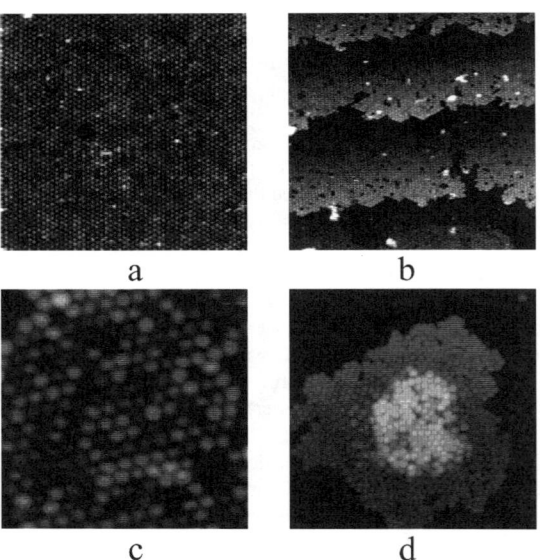

Figure 9. *Virus particles appear to have different sizes in an STMV cubic crystal (a) and in a BMV crystal (b) as well. STMV particles from cubic crystals dried on a mica surface and imaged in propanol (c). STMV particles from a solution which produces orthorhombic crystals, dried and imaged in propanol (d). Scan sizes are 1 x 1 μm² (a), 2 x 2 μm² (b), 260 x 260 nm² (c) and 610 x 610 nm² (d)*

ACKNOWLEDGMENTS

The authors would like to thank Steven B. Larson for discussions and review of the manuscript, John Day and Jiashu Zhou for preparation of STMV solutions and Aaron Greenwood and Debora Felix for technical support. Research supported by the National Aeronautics and Space Administration.

REFERENCES

1. J. M. Wiencek and W. F. Jones, Abstracts of the 8th International Conference on the Crystallization of Biological Macromolecules, May 14-19, 2000, Sandestin, Florida (USA), p. 40.

2. A. McPherson, *Crystallogr. Rev.* **6**, 157-308 (1996).

3. B. R. Thomas, P. G. Vekilov and F. Rosenberger, *Acta Cryst.* Sec D, **52**, 776-789 (1996).

4. C. L. Caylor, I. Dobrianov, S. G. Lemay, C. Kimmer, S. Kriminski, K. D. Finkelstein, W. Zipfel, W.W. Webb, B. R. Tomas, A. A. Chernov, R. E. Thorn, *Proteins-Struct. Func. and Gen.* **36**, 270-281 (1999)

5. R. A. Valverde, J. A. Dodds, *J. Gen. Virol.* **68**, 965-972 (1987).

6. S. B. Larson, S. Koszelak, J. Day, A. Greenwood, J. A. Dodds, A. McPherson, *Nature* **361**, 179 (1993).

7. S. B. Larson, J. Day, A. Greenwood and A. McPherson, *J. Mol. Biol.* **277**, 37-59 (1998).

8. J. Israelachvili, *Intermolecular & Surface Forces,* 2nd ed. (Academic Press, San Diego, 1992), p.246.

9. A. J. Malkin, Yu. G. Kuznetsov, and A. McPherson, *J. Struct. Biol.* **117**, 124-137 (1996).

Mat. Res. Soc. Symp. Vol. 620 © 2000 Materials Research Society

Macromolecular Crystals in the Service of Biotechnology and Medicine

Alexander McPherson
University of California, Irvine
Dept. of Molecular Biology and Biochemistry
Irvine, CA 92697

ABSTRACT

The application of genetic and molecular biological methods to the engineering of proteins, the engine of the biotechnology revolution, has become increasingly dependent on knowledge of protein, nucleic acid and virus molecular structure. Structural information of high precision, essential to this enterprise, can only be obtained through X-ray crystallographic techniques. The quality of this information is directly a function of the properties and degree of perfection of the crucial intermediates, the macromolecular crystals. As a consequence, there is now broad and intense interest in developing new methods, procedures and reagents for the nucleation and growth of such crystals. There is, in addition, an expanded interest in the properties of biological crystals and the use of physical measurements in improving approaches to growing better and larger crystals. Here some of the fundamentals of macromolecular crystal growth will be reviewed, and some current trends in the field remarked upon, including the new initiative to develop a national structural genomics program.

INTRODUCTION

Progress in molecular biology and its application to human medicine, agriculture, and industrial processes have for the past two decades been crucially dependent on knowledge of macromolecular structure at the atomic level, and this has included proteins, nucleic acids, viruses, and other large macromolecular complexes and assemblies. Redundancies of structural elements now emerging from more recently determined molecular structures suggests that the number of structural motifs and substructures (domains) naturally occurring may be finite, and that ultimately all macromolecular structures may be classified and catalogued according to polypeptide folds. Once all, or most of the folds which are utilized by nature are known, then this will provide predictive insight, based primarily on amino acid sequence, of the structures and functions of unknown proteins. The sequences of most proteins, it is important to note, are currently being defined by a broad array of sequencing efforts, such as the human genome project, carried out by both government and the private sector. Extension of these genome projects to the three dimensional structural level appears the next logical step, and this effort, under the broad rubric of structural genomics, is now in the planning stages.

In addition to the dramatic impact that knowledge of three dimensional structures of proteins has had on fundamental research in biochemistry and biology, macromolecular structure is of formidable value in biotechnology as well. Here, it provides the essential knowledge required to apply the technique of rational drug design in the creation and discovery of new drugs and pharmaceutical products (McPherson, 1992, 1994). It serves as the basis of powerful approaches now being applied in small, emerging biotechnology companies as well as major pharmaceutical companies to identify lead compounds to treat a host of human ailments, veterinary problems, and crop diseases in agriculture. The underlying hypothesis is that if the structure of the active site of a salient enzyme in a metabolic or regulatory pathway is known, then chemical compounds, such as drugs, can be rationally designed to inhibit or otherwise affect the behavior of that enzyme.

A second approach, of equal importance to biotechnology that also requires knowledge of three-dimensional macromolecular structure is the genetic engineering of proteins. Although recombinant DNA techniques provide the essential synthetic role that permits modification of the proteins, structure determination provides the analytical function. It serves as a structural guide for intelligent and purposeful changes, in place of random and chance amino acid substitutions. Direct visualization of the structural alterations that are introduced by mutation offers new directions for chemical and physical enhancements.

Presently, and in the foreseeable future, the only technique that can yield atomic level structural images of biological macromolecules is X-ray diffraction analysis as applied to single crystals. While other methods may produce important structural and dynamic data, for the purposes described above, only X-ray crystallography is adequate. As its name suggests, application of X-ray crystallography is absolutely dependent on crystals of the macromolecule, and not simply crystals, but crystals of sufficient size and quality to permit accurate data collection. The quality of the final structural image is directly determined by the perfection, size and physical properties of the crystalline specimen, hence the crystal becomes the keystone element of the entire process, and the ultimate determinant of its success (McPherson, 1989).

When crystallizing proteins for X-ray diffraction analysis, one is usually dealing with pure, often exceptionally pure macromolecules, and the objective is to grow only a few large, perfect crystals. It is important to emphasize that while the number of crystals needed may be few, often the amount of protein available may be severely limited. This in turn places grave constraints on the approaches and strategies that may be used to obtain those crystals. While new methodologies such as synchrotron radiation (Helliwell, 1992) and cryocrystallography (Garmen and Schneider, 1997) have driven the necessary size of specimen crystals consistently downward, they have not alleviated the need for crystal perfection. The X-ray analysis is a singular event confined to the research laboratory and the final product is basic scientific knowledge. The crystals themselves have no medicinal or pharmaceutical value, but simply serve as intermediaries in the crystallography process. The crystals provide the X-ray diffraction patterns that in turn serve as raw data allowing the visualization of the macromolecules composing the crystals.

SOLUBILITY AND SUPERSATURATION

Crystallization of a macromolecule requires the creation of a supersaturated state. This is a non-equilibrium condition in which some quantity of the macromolecule in excess of the solubility limit, under specific chemical and physical conditions is, nonetheless, present in solution. Equilibrium is re-established by formation and development of a solid state, such as crystals, until the saturation limit is attainted. To produce a supersaturated solution, the properties of an undersaturated solution must be modified to reduce the ability of the liquid to solubilize the macromolecule, or some property of the macromolecules must be altered to reduce their solubility and/or to increase the attraction of one macromolecule for another. In all cases, the relationships between solvent and solute, or between the macromolecules in solution, are perturbed so as to inspire formation of the solid state.

There are distinctive differences between the crystallization of macromolecules and conventional, small molecules (Feigelson, 1988; Feher, 1986; Durbin and Feher, 1996; McPherson, 1989; McPherson, et al 1995). First, macromolecules may exist in several solid states that include amorphous precipitates as well as crystals. Second, macromolecular crystals nucleate, or initiate development, only at very high levels of supersaturation, often two to three orders of magnitude higher than are required to sustain growth. Finally, the kinetics of macromolecular crystal nucleation and growth are generally two to three orders of magnitude slower than for conventional molecules (Kuznetsov, et al 1995; Malkin et al 1997,1996). This latter difference arises from the considerably larger size, lowered diffusivity, and weaker association tendencies compared with small molecules or ions, as well as a lower probability of incorporation of an incoming macromolecule into a growth step.

PRECIPITANTS AND THEIR EFFECTS

The solubility of a macromolecule in concentrated salt solutions is complicated, but it can be viewed naively as a competition, between salt ions, principally the anions, and the macromolecules, for the binding of water molecules, which are essential for the maintenance of solubility (Hofmeister, 1890; Herriott, 1942; Cohn and Edsall, 1943; Cohn and Ferry, 1943). At sufficiently high salt concentrations the macromolecules become so uncomfortably deprived of solvent that they seek association with one another in order to satisfy their electrostatic requirements. In this environment ordered crystals, as well as disordered, amorphous precipitate may form. Some salt ions, chiefly cations, are also necessary to insure macromolecular solubility. At very low ionic strengths, cation availability is insufficient to maintain macromolecule solubility, and under those conditions too, crystals may form.

Organic solvents reduce the dielectric of the medium, hence the screening of the electric fields that mediate macromolecular interactions in solution. As the concentration of organic solvent is increased, attraction between macromolecules increases, solvent becomes less effective (the activity coefficient of water is reduced) and the solid state is favored (Cohn, et al., 1974; England and Seifter, 1990).

Some polymers, polyethylene glycols are most prominently employed, produce volume exclusion effects that also induce separation of macromolecules from solution Ingham, 1990; McPherson, 1976). The polymeric precipitants writhe and twist randomly in solution and, like anti social individuals with wildly swinging elbows, occupy far more space than they otherwise deserve. This results in less available space for the other macromolecules which then segregate into aggregates, and ultimately form a solid state, often crystals.

Increase in the concentration of precipitants, salts, organic or polymeric, provides one method for macromolecule crystallization, but there are other approaches (see McPherson, 1982; 1990; 1998; Ducruix and Giegé, 1992; Durbin and Feher, 1996). The solubility of some macromolecules is affected by temperature, and its manipulation may be employed to produce supersaturated solutions. Virtually all macromolecules are sensitive, in terms of their physical-chemical behavior, to the pH, or hydrogen ion concentration of their solution. This arises from the charged and ionizable nature of amino acid groups on their surfaces. Just as solubility may be reduced by the addition of precipitating agents, it may be manipulated as well by suitable alterations in the pH of the solution.

CREATING SUPERSATURATION – METHODOLOGY

Changes in solution properties may be affected by a wide range of techniques, all designed to transform solutions from undersaturated to supersaturated states (McPherson, 1982, 1989, 1998; Ducruix and Giegé, 1992). These include simple, straightforward approaches such as the direct addition of precipitating agents, acids, or bases to macromolecular solutions (batch crystallization), the diffusion of the components of one solution into another (free interface or liquid-liquid diffusion), and more sophisticated techniques where change is brought about through the vapor phase (vapor diffusion methods).

A traditional and still popular method for altering solution properties through the liquid phase is dialysis. This is essentially the same as free interface diffusion except that a semi permeable membrane is inserted between the protein and precipitant phases. This has the effect of immobilizing the macromolecular solution while allowing only buffer ions and precipitant molecules to equilibrate. Dialysis can also be carried out on a microscale using plastic dialysis buttons or small tubes.

Virtually all approaches currently in use employ amounts of protein solution in the 1- 10 µl range and therefore require micropipettes and micro techniques. Plasticware and other devices for carrying out procedures on these small scales are commercially available and readily obtained. Among the numerous approaches, vapor diffusion is most commonly used, particularly to identify initial crystallization conditions. Vapor diffusion may be carried out using "sitting drops" where the microdroplet is deployed atop some surface such as a glass plate or slide, or in the "hanging drop" configuration, where it is suspended from a glass or plastic slip (McPherson, 1982, 1998). In both cases the droplet is exposed to a reservoir solution much greater in volume and at higher precipitant concentration. By loss of solvent through the vapor phase, conditions in the protein microdrop are altered, a supersaturated state is created, and crystallization promoted.

FACTORS AFFECTING CRYSTALLIZATION

There are many factors that affect the crystallization of macromolecules (McPherson, 1982, 1998). These may affect the probability of its occurrence at all, nucleation probability and rate, crystal growth rate, and the ultimate sizes and quality of the products. As noted above, pH and salt, or the concentrations of other precipitants are of great importance. The concentration of the macromolecule, which may vary from as low as 2 mg/ml to as much as 100 mg/ml, is an additional, significant variable.

Other parameters may be less important but may play crucial roles. The presence or absence of ligands or inhibitors, the variety of salt or buffer, the equilibration technique used, the temperature, or the presence of detergents, these are all pertinent considerations. Parameters of somewhat lesser significance are things like gravity, electric and magnetic fields, or viscosity. It can, in general, not be predicted which of these many variables may be of importance for a particular macromolecule, and the influence of any one must, in general, be defined by a series of empirical trials.

A factor of particular importance is the purity of the macromolecule (Giegé, et al, 1994) and this deserves special emphasis. Some proteins, it is true, may crystallize even from very impure mixtures, and indeed, crystallization has long been used as a powerful purification tool. In general, however, the likelihood of success in crystal growth is greatly advanced by increased homogeneity of the sample. Investment in further purification is always warranted, and usually profitable. When every effort to crystallize a macromolecule fails, the best recourse is to further purify.

SCREENING AND OPTIMIZATION

When attempting the crystallization of a macromolecule, there are two objectives that must be reached. The first is to identify initial conditions, which yield some kind of crystals, even if they are too small or misshapen for effective diffraction analysis. The second is, using these as starting points, to optimize the crystal growth by varying all relevant parameters to obtain crystals adequate for X-ray structure determination.

There are two approaches to identifying crystallization conditions, and these are referred to as sparse-matrix screens (Carter and Carter, 1979; Jancarik and Kim, 1991), and systematic screens. In both cases, commercially available kits are frequently used. Sparse-matrix kits usually contain twenty-four or forty-eight (corresponding to the number of wells or chambers in most crystallization plates) crystallization solutions, which sample a broad range of different precipitants, ions, buffers, pH, or additives. The specific conditions represented by each of the solutions reflects experience acquired over the years, and try to select those conditions which have often proven successful in the past. The intervals of sampling are, however, very coarse and this may result in overlooking conditions that might yield crystals.

The second approach (McPherson, 1982, 1989, 1998) is to use a finer sampling of conditions for a specific precipitant, and then repeating the trials with another precipitant type, or in the presence of a ligand which might affect the physical-chemical behavior of the macromolecule. With this systematic strategy, a given precipitant may be used at a regularly increasing set of concentrations and at a varying sequence of pHs. This produces a two dimensional matrix of crystallization conditions which is also suitable for use with most available crystallization plates. The major disadvantage of this approach, though it is clearly more thorough, is that it requires more material. If the amount of macromolecule is severely limiting, then some intelligent compromise of the sparse-matrix and systematic screen may have to be designed based on the known solubility properties of the macromolecule.

In the process of optimization, the variables of precipitant concentration, pH, temperature, etc. are expanded into a finely sampled, multi-dimensional matrix of conditions centered on the initial condition, which produced crystals. In some cases, initial conditions may not be further improved sufficiently, hence it is important to continue screening for other successful conditions even while attempting to optimize the first. In addition to chemical and biochemical variables, it is also necessary to explore different crystallization methods, and even different geometries of the same system, e.g. both hanging and sitting drops with vapor diffusion.

While a considerable degree of success, compared to previous decades, has been achieved in recent times in the area of macromolecular crystallization, and many new structures are rapidly appearing, some major problems remain. As a consequence, the crystallization of a protein still is the primary obstacle to structure determination. This is not to say that the X-ray crystallographic process is not still arduous, intellectually exacting, and demands skill and experience, but that once respectable crystals of a macromolecule are obtained, it is a reasonable expectation that its structure will eventually be determined. This is particularly true of some major classes of proteins, such as the immunoglobulins and highly dynamic and mobile multidomain molecules, lipophilic molecules, molecules which aggregate badly or are membrane associated, large multimeric complexes, and many small proteins and polypeptides that include important hormones. Should the effort to implement a structural genomics enterprise be undertaken, then many of the gene products will fall into these categories, and they will be problematic indeed. Any effort at mass throughput of crystal structures will unquestionably encounter the obstacle of crystallization. To be successful, this enterprise must have methods for rapid and assured crystallization of the target macromolecules.

The crystallization of any molecule involves two contiguous stages, nucleation and growth. The first is characterized by the formation, according to statistical probability based on the thermodynamic properties of the system, of a "critical nucleus." This is defined as an ordered aggregate of molecules, which attracts new molecules to its bulk at a higher rate than molecules depart. Once such a critical size is attained, the properties of the ultimate crystal are determined by the manner in which new molecules are recruited into the surfaces of the growing crystals. Molecules may be incorporated by a variety of growth mechanisms (Malkin, et al 1995) and the kinetics of the various processes are dependent on the intermolecular interactions, transport processes operating in solution, the degree of supersaturation, and the types and concentrations of the impurities present in the system. While we have a reasonable, but admittedly still incomplete understanding of the growth processes and how they may be controlled, we have very little appreciation of the mechanisms by which critical nuclei appear. This is unfortunate, because without nucleation there are no crystals at all and X-ray crystallography is out of the question.

In the growth of small, conventional molecules, the problem of nucleation is frequently obviated by using seed crystals of the material under study, and in those cases the problems of crystallization arise principally from control of growth mechanism and growth kinetics. This is generally not possible with macromolecules, because no preexisting crystals are available. Usually, one is attempting to crystallize the protein for the first time, hence nucleation is of particular significance. Experience from both conventional, and now macromolecular crystallization has, however, suggested some alternatives. In particular, surfaces in contact with the mother liquor from which the crystals form are now recognized to be of substantial importance in accelerating the nucleation process. Molecules, and especially macromolecules having extensive surfaces with diverse charge distributions, adsorb to, and are concentrated at surfaces in contact with their solutions. In addition, many surfaces, because of their own charge properties and microstructure help to organize the macromolecules into more ordered assemblies and arrays. These then serve to promote the formation of critical nuclei. In a sense, the foreign surfaces tend to act as catalysts to lower the activation energy for nucleation and increase the probability that nuclei will form and the crystallization process begin.

There are, in fact, several problems that must be addressed if the crystallization of proteins is an integral part of a serious structural genomics program, and even if it is only to keep up with the current demand for macromolecular structure by the biochemical and molecular biological research community. The process of crystallization, which must for now remain fundamentally an empirical endeavor, must be (1) made more automated so that more proteins and more plausible sets of conditions can be addressed in parallel, (2) crystallization trials and screening matrices must be put on a micro scale so that more conditions can be explored for a given quantity of protein, since protein will frequently be only sparingly available, and (3) new ways must be discovered to overcome the nucleation problem so that the first step in crystallization, the most difficult step in terms of thermodynamics, kinetics, and practical considerations, can be more readily overcome.

FUTURE TRENDS

In terms of basic research, our goal must be to dramatically accelerate the rate at which new protein and indeed new nucleic acid and virus structures are determined and made available to the molecular biologists and biochemists of the world. Currently, our understanding of biological function and its relationship to molecular structure is severely limiting and this arises directly from our limited ability to obtain X-ray quality crystals of important biological macromolecules. Expanding the limits of crystallization will produce a corresponding extension of the limits of molecular biological knowledge. Advances in medicine, treatment of disease, compensation of deficiencies, correction of defects all are now tied directly to our progress in molecular biology. Antibody engineering, synthetic blood, proteinaceous drugs, and a host of other novel approaches to pharmacology and medicine can only achieve their ultimate promise is there is an adequate structural basis. In a very practical sense, this means an adequate supply of protein crystals.

To emphasize this point, we can consider the example of an extremely important, novel, and highly visible program now emerging as a natural extension of the human genome project, the enterprise of Structural Genomics. This enterprise, likely to involve thousands of molecular and structural biologists has as its primary objective the three-dimensional structure determination of all of the gene products of the human genome. Logically, this effort will not even be confined to the human genome, but will ultimately include most of the organisms that populate the living world. Such an endeavor will require the cloning, expression, crystallization, and X-ray structure determination of hundreds of thousands of unique protein molecules. While the genetic components of the problem and the structure determination aspects are now relatively well in hand, crystallization remains an ominous and forbidding obstacle. Compared to the other parts of the problem, the available technologies for protein crystallization are relatively primitive. The success of the entire structural genomics project may well depend, at end of day, on the advancement of macromolecular crystallization technology.

Just as the structural genomics enterprise will be limited by protein crystallization, advances in the biotechnology industry, likely to be the foremost American industry in the early part of the twenty first century, will be similarly impeded. Breakthroughs in the private sector application of molecular biology will, like structural genomics, be critically dependent on progress in devising more assured and expeditious approaches to macromolecular crystallization. This applies not only to the medical profession and the pharmaceutical industry, but to other sectors as well. These include for example agriculture where new, and environmentally acceptable insecticides and herbicides will be needed, and in chemical manufacturing where engineered enzymes will increasingly be employed to catalyze unfavorable reactions.

Increasing efficiency and speed, and making protein crystallization available to an ever-expanding community will also serve to introduce the benefits of macromolecular structure to new sectors of the economy, and to new fields of scientific research. Among these are the newly developing fields of biomaterials and biocomposits, fields that may produce, based on biological structure, whole new classes of materials. And with the new materials, as bronze succeeded stone and iron triumphed over bronze, entire new possibilities for human endeavor.

ACKNOWLEDGEMENTS

The author wishes to thank his colleagues, Drs. Alexander Malkin and Dr. Yurii G. Kuznetsov for stimulating and educational discussions. Research featured here was funded by the National Institutes of Health and by the National Aeronautics Space Administration.

REFERENCES

1. A. McPherson, Preparation and analysis of protein crystals, (John Wiley, 1982) pp. 371.
2. A. McPherson, The Role of Protein Crystals in Biotechnology and Industry in, *Proceedings of Frontiers in Bioprocessing II*, (American Chemical Society Boulder, CO, 1992) pp. 10-15.
3. A. McPherson, The role of X-ray crystallography in structure based rational drug design, *Rational Drug Design* (CRC Press, 1994) Chptr. 6, p 170.
4. A. McPherson, *Scientific American*, **260 (3)** 62-69 (1989).
5. A. McPherson, *Eur. J. Biochem.* **189** (1), 1-23 (1990).
6. E. F. Garman, and T. R. Schneider, *J. Appl. Cryst.* **30**, 211-237 (1997).
7. G. Feher, *J. Cryst. Growth* **76**, 545-546(1986).
8. R. S. Feigelson, *J. Cryst. Growth* **90**, 1-13 (1988).
9. S. D. Durbin and G. Feher, *Annu. Rev. Phys. Chem.* **47**, 171-204 (1996).
10. J. R. Helliwell, *Macromolecular Crystallography with Synchrotron Radiation.* (Cambridge University Press, 1992).
11. A. McPherson, A. J. Malkin, and Yu. G. Kuznetsov, *Structure* **3**, (8), 759-768 (1995).
12. T. Hofmeister, *Physiol. Chem.* **14**, 165 (1890).
13. E. J. Cohn, and J. D. Ferry in, *Proteins, Amino Acids and Peptides*, eds. E. J. Cohn and J. T. Edsall (Van Nostrand-Rheinhold, New Jersey 1943) p. 586.
14. E. J. Cohn, and J. T. Edsall in, *Proteins, amino acids and peptides as ions and dipolar ions*, eds. E. J. Cohn and J. T. Edsall, (Van Nostrand-Rheinhold, New Jersey 1943) p. 586.
15. R. M. Herriot, *Chem. Rev.* **30**, 413 (1942).
16. E. J. Cohn, W. L. Hughes, and J. H. Weare, *J. Am. Chem. Soc.* **69**, 1753-1761 (1947).
17. S. Engard, and S. Seifter, *Methods in Enzymology* **182**, 301-306 (1990).
18. K. C. Ingham, *Methods in Enzymology* **182**, 301 (1990).
19. A. Ducruix, A. and R. Giége, Crystallization of Nucleic Acids and Proteins, A Practical Approach (IRL Press, Oxford 1992).
20. Yu. G.. Kuznetsov, A. J. Malkin, A. Greenwood and A. McPherson. *J. Struct. Biol.* **114(3)**, 184-196 (1995).
21. A. J. Malkin, Yu. G. Kuznetsov and A. McPherson. *Surface Science* **393**, 95-107 (1997).
22. A. J. Malkin, Yu. G. Kuznetsov, W. Glantz and A. McPherson. *J. Phys. Chem.* **100**, 11736-11743 (1996).

Inorganic Systems II— Impurities and Defects

Mat. Res. Soc. Symp. Vol. 620 © 2000 Materials Research Society

Moving Steps and Crystal Defects on Spinel Surfaces

S.V. YANINA and C.B. CARTER
Department of Chemical Engineering and Materials Science, University of Minnesota,
Minneapolis, MN 55455

ABSTRACT

The morphology of the reconstructed $\{001\}$ surface of $MgAl_2O_4$ spinel is studied by scanning probe microscopy (SPM). The observations show that the $\{001\}$ surface of $MgAl_2O_4$ may exist as two variants which are related through 90° rotation about the [001] axis. These surface variants exhibit different lateral forces and tend to grow/evaporate and to etch anisotropically along either the [110] or the [1$\bar{1}$0] directions of the crystal. Surface terraces that are formed by different variants were found to be separated by ~2.0 Å-high steps, while the terraces which belong to the same variant are separated by ~4.0 Å-high steps. It is expected that the origins of the preferential motion of ledges on the $\{001\}$ spinel surface is related to the anisotropic distribution of cations along either the [110] or the [1$\bar{1}$0] directions within the $\{001\}$ crystal planes in the spinel crystal.

INTRODUCTION

Many technologically important applications of ceramic materials, such as their use as catalysts, in information storage systems, etc., depend on effective utilization of certain structural, chemical and mechanical properties of their surfaces [1]. Complex oxides, such as ferrites and transition-metal-based spinels, are widely used in industry. In order to understand better their surface properties, processes such as step formation and dynamics of step motion, need to be examined. The objective of the present project is to study surface dynamics of the prototypical spinel, $MgAl_2O_4$, and to investigate high-temperature interactions between moving surface steps and lattice defects, such as dislocations, which are present in a crystal.

The structure of the $MgAl_2O_4$ crystal (space group symmetry Fd$\bar{3}$m) [2] may be thought of as being created by alternating $\{008\}$ planes filled by tetrahedrally-coordinated Mg^{2+} cations with $\{004\}$ planes formed by O^{2-} cations together with Al^{3+} cations in octahedral coordination (see Figure 1). In spinel structure, there are four $\{008\}$ and four $\{004\}$ planes per unit cell, with the distance separating the nearest-neighbor pair of $\{004\}$ (or $\{008\}$) planes being equal to ~2.1 Å, or a/4[001]. Any pair of adjacent $\{004\}$ (or $\{008\}$) planes is fully equivalent chemically, and differs only by the way of cation arrangement. If in one plane, the cations fill the crystal along the [110] direction, then, in the other plane, the cation-filled direction is the [1$\bar{1}$0]. A pair of $\{004\}$ (or $\{008\}$) planes separated by ~4.2 Å, or a/2[001], is fully equivalent, both chemically and structurally. It is expected that the $\{001\}$ surface of spinel may reflect non-equivalency of different crystal planes of $\{001\}$ orientation with respect to a 90°-rotation about the [001] crystal axis.

EXPERIMENTAL

2×2×1 mm samples of single-crystal $MgAl_2O_4$ spinel of the (001) orientation were polished and cleaned in *aqua regia*, acetone and methanol. Selected samples were subjected to acid-etching in *aqua regia* for the duration of 24 hours (at room temperature). Heat-treatments of the samples were performed in a sintered spinel crucible at a pressure of 10^{-4}–10^{-5} Torr in a Centorr™ furnace at 1200–1800°C for 8 hours. Surface crystallographic orientations were checked by Laue backscattered X-ray diffraction. The chemical composition of the surface was determined *ex situ* by X-ray energy-dispersive spectroscopy (XEDS) and X-ray photoemission spectroscopy (XPS). SPM images were collected in air in contact mode of operation on a Nanoscope III (Digital Instruments, Santa Barbara, CA). Cantilevers used in present study were V-shaped Si_3N_4 cantilevers (Ultralevers, Park Inst., Sunnyvale, CA) with a nominal spring constant of 0.12 N/m.

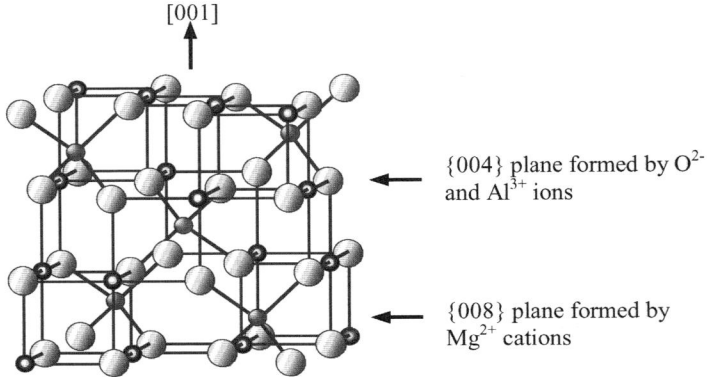

[001]

← {004} plane formed by O^{2-} and Al^{3+} ions

← {008} plane formed by Mg^{2+} cations

Figure 1. Normal spinel structure. Oxygen ions are shown as large gray spheres, cations in tetrahedral coordination as small gray spheres, and octahedrally coordinated cations as small dark spheres. The distance between the two nearest-neighbor pair of the {004} or the {008} planes is ~2.1 Å

RESULTS

The heat-treatment of polished slightly vicinal surfaces of $MgAl_2O_4$(001) in the 1200–1600°C temperature interval leads to their reconstruction into terrace-and-step morphology (see Figure 2(a)). The steps align preferentially along [110] and [1$\bar{1}$0] directions of the crystal. The average height of the smallest step found on the {001} surface is ~2.0 Å which is close to the inter-planar distance for the {004} (or {008}) planes in spinel structure (~2.1 Å, or a/4[001]). The average height of the typical ledge on the {001} surface of spinel which was heat-treated at 1200°C is ~10–100 nm. The average surface area of the terrace depends on the degree of misalignment between the direction of the polished plane and the position of the {001} planes in the crystal. Combined acid-etching and thermal etching of the {001} surface of $MgAl_2O_4$ (see

Figure 2(b)) also proceeds preferentially along the [110] and the [1$\bar{1}$0] directions of the crystal. Such etching results in formation of narrow elongated trenches on the crystal surface. For each surface terrace, there is only one preferred direction of trench formation, either along the [110] or the [1$\bar{1}$0] direction.

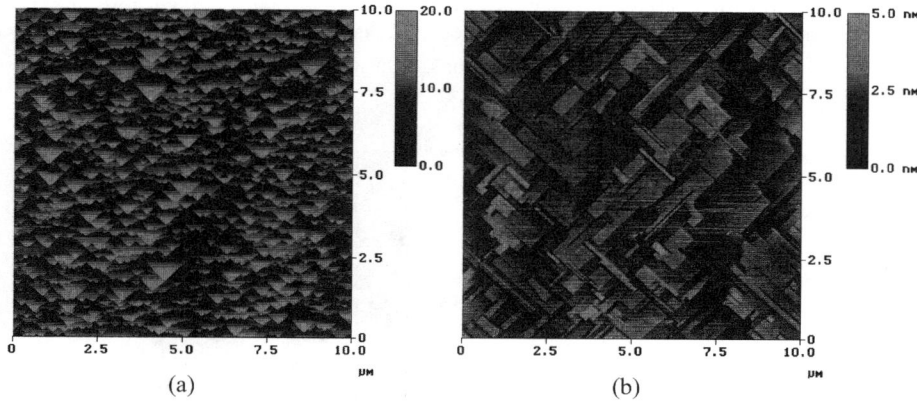

Figure 2. Height-mode SPM image of the annealed {001} surfaces of MgAl$_2$O$_4$. (a). The as-polished surface was annealed at 1200°C for 8 hours. The ledges on the surface are aligned along the [110] and [1$\bar{1}$0] directions of the crystal. (b). The surface was etched at room temperature by *aqua regia* for the duration of 24 hours and subsequently heat-treated at 1200°C for 8 hours. The trenches in the surface are aligned along the [110] and [1$\bar{1}$0] directions of the crystal.

It is proposed that the existence of preferential directions of terrace growth and etching of the {001} surface of MgAl$_2$O$_4$ is closely related to the intrinsic anisotropy of the {001} crystal planes in spinel structure. As discussed above, a pair of adjacent {004} (or {008}) planes is related through the 90° rotation about the [001] axis. Since the preferred direction of cation distribution is the only feature that is different for such a pair of planes, it is expected that, provided they form stable {001} surface terminations in a spinel crystal [3], such planes will have identical surface energies. This should hold true even in the event of significant reconstruction of the MgAl$_2$O$_4$(001) surface, as, from electrostatic and chemical considerations [4], the mechanism of surface reconstruction will be the same for both planes. If constructed from such planes, the {001} surface of spinel may be considered as consisting of two different surface domains of two-fold symmetry which separated by anti-phase domain boundaries [5], or ~2.0 Å-high surface steps.

It should be mentioned that mixed pairs of planes in spinel, such as the {008}/{004} pair, have different chemical composition and are separated by ~1.0 Å in the bulk crystal. Since no steps of ~1.0 Å height was detected on the reconstructed spinel surfaces, the stable {001} surface terminations of MgAl$_2$O$_4$ are produced by only one type of crystal planes (either of the {004} or of the {008} orientation).

Differences in the symmetric properties of the two variants of the surface structure of MgAl$_2$O$_4$(001) become apparent when the height-mode SPM images are compared with the lateral-force SPM images of the same surface areas. Figure 3(a) shows, that while the surface terraces that belong to one of the variants exhibit high friction forces when scanned along the [010] direction of the crystal, the terraces that are formed by the other surface variant appear dark (show low friction forces). When the direction of scan is rotated through 180°, the contrast is reversed. Remarkably, when the lateral-force image is collected with the direction of scan being parallel to either [110] or [1$\bar{1}$0] directions, both surface variants demonstrate friction forces of the same magnitude.

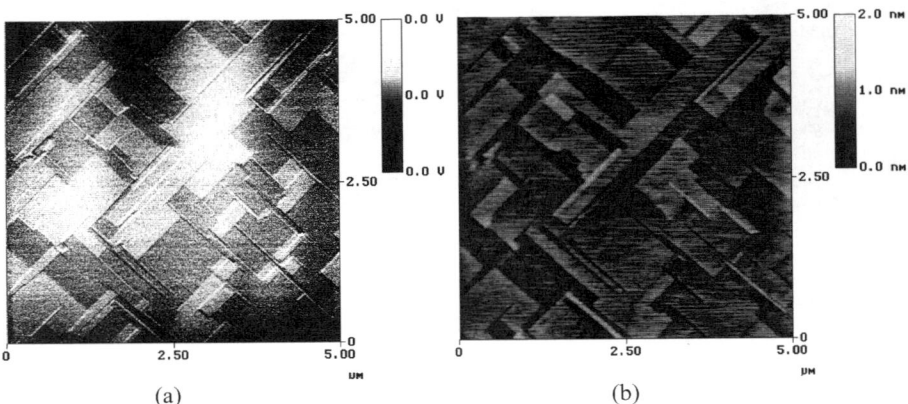

(a)   (b)

Figure 3. SPM images of the {001} surface of MgAl$_2$O$_4$ which was acid-treated and annealed at 1200°C for 8 hours. (a). Lateral-force image. Two different surface domains coexisting on the MgAl$_2$O$_4$(001) show different contrast: the terraces that belong to the first symmetry domain are bright, while the terraces that are formed by the second domain are dark. (b). Height-mode image of the same surface area. Surface terraces are separated by nanoledges aligned along the [110] and the [1$\bar{1}$0] directions of the crystal.

Growth/evaporation anisotropy of surface variants on the MgAl$_2$O$_4$(001) is especially apparent during the processes of thermal etching. Figure 4(a) shows a dislocation termination site on the {001} surface which was heat-treated at 1500°C for 8 hours. The height of a half-ledge created at the dislocation termination site [6] is ~4.0 Å, which is close to a/2[001]. a/2[110]is the normal to the {001} surface component of the shortest Burgers vector in spinel structure, a/2[011] [7]. The parallel surface steps seen on the image are aligned along the [110] direction of the crystal and separate terraces which belong to the same surface variant.
At higher temperatures, when the motion of surface steps is fast [8], the thermal etching of the dislocation termination sites on the MgAl$_2$O$_4$(001) produces two, instead of one, half-ledges (see Figure 4(b)). These half-ledges which are ~2.0 Å, or a/4[001], high, are orthogonal to each other and are aligned along the [110] and the [1$\bar{1}$0] directions of the crystal. Combined, their heights sum up to ~4.0 Å, or a/2[001]. Rotation of these half-ledges about the dislocation termination

site creates a topologically complex evaporation pattern [9] which consists of two groups of steps. The long steps that form evaporation pattern sides are ~4.0 Å-high, while the short steps which can be seen in its corners are ~2.0 Å high. The ~4.0 Å steps separate terraces that belong to the same surface domain; the ~2.0 Å steps adjoin terraces which belong to the different surface domains. The steps are aligned along the [110] and the [1$\bar{1}$0] directions of the crystal.

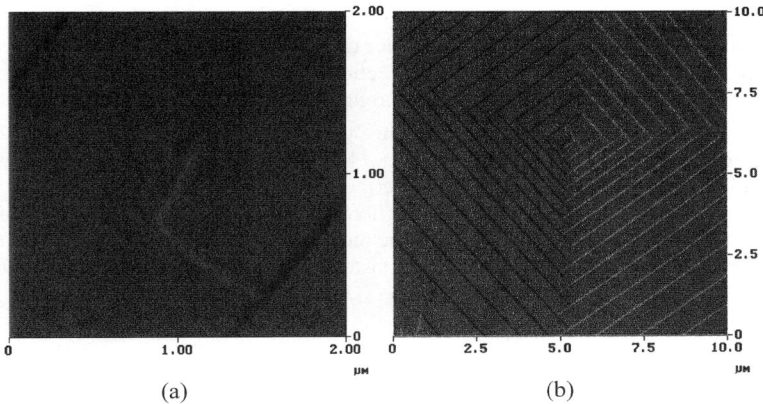

(a) (b)

Figure 4. Deflection-mode SPM images of thermally-etched dislocation termination sites on the {001} surface of MgAl$_2$O$_4$. Direction of scanning was from left to right. (a). After heat-treatment at 1500°C for 8 hours. Single ledge created at the dislocation termination site is ~4 Å high. Upon emergence, the ledge rotates by 180° and merges with the nearby surface step. (b). After heat-treatment at 1800°C for 8 hours. A pair of ledges at spiral origin at the dislocation termination site are ~2 Å high. In the course of evaporation, the ledges rotate about the dislocation termination point with 90° "phase shift". Surface steps are aligned along the [110] and the [1$\bar{1}$0] directions of the crystal.

It is expected that the formation of two, instead of one, ledges at the dislocation emergence sites on the {001} surface of spinel is related to the fact that the development of variants on the {001} surface of spinel, which occurs during surface reconstruction, chemical etching or evaporation, proceeds anisotropically. While one surface variant grows through the propagation of steps aligned along the [110] direction, the other variant evolves through motion of steps which are aligned along the [1$\bar{1}$0] direction. Formation of two half-ledges at the dislocation termination site may be understood as follows. The original ~4.0 Å-high half-ledge is fixed at the dislocation termination point. Evaporation from this half-ledge would ordinarily be expected to result in its rotation about the point of its origin (i.e., the dislocation termination site [6]). In contrast, its anisotropic evaporation proceeds in two steps. At first, the removal of material along the [110] direction (i.e., perpendicular to the direction of ledge alignment) creates two parallel half ledges which are ~2.0 Å high. Such ledges separate two different surface variants and are expected to have orthogonal preferred directions of alignment. One of these ~2.0 Å ledges tends to align along the [1$\bar{1}$0] direction, but, when formed as a result of evaporation of the original

~4.0 Å-high ledge, is created with the unfavorable direction of alignment along the [110] direction. In order to attain its preferred alignment, this ledge starts rapid motion. This creates a pair of perpendicular ~2.0 Å ledges at the dislocation termination site. In the course of further evaporation, these ledges continue rotation with the 90° "phase difference" between them [9].

CONCLUSIONS

In the present paper, dynamics of surface development of $MgAl_2O_4(001)$ is discussed. It is proposed that the high-temperature reconstruction of the {001} surface of spinel proceeds *via* formation of two domains which are related through 90° rotation about the [001] axis. Such domains tend to grow/evaporate and to etch anisotropically along the [110] or the [1$\bar{1}$0] directions of the crystal. On the {001} surface of $MgAl_2O_4$, the ~2.0 Å-high steps serve as anti-phase domain boundaries separating surface terraces that belong to different domains. The observations indicate that the two surface domains are fully equivalent in terms of surface energies and growth rates, and vary only by the preferred directions of growth and by their friction force coefficients. It is expected that existence of surface domains is due to the preferential arrangement of cations along the [110] and the [1$\bar{1}$0] directions in the crystal planes of {001} orientation.

ACKNOWLEDGEMENTS

SVY is in the Chemical Physics Program at the University of Minnesota. This Research has been supported by the U. S. Department of Energy under Grant No DE-FG02-92ER45465. The SPM used is part of the IT Characterization Facility at the University of Minnesota.

REFERENCES
[1] M. Naoe, N. Matsushita, K. Noma, and S. Nakagawa, J. de Physique IV **7**, C1-735-738 (1997).
[2] P. Villars and L.D. Calvert, *Pearson's Handbook of Crystallographic Data for Intermetallic Phases, American Society for Metals,* American Society for Metals, Metals Park, **1989**.
[3] P.W. Tasker, J. Phys. C **12**, 4977 (1979).
[4] D. Wolf, Phys. Rev. Lett. **68**, 3315 (1992).
[5] G. Tarrach, D. Bürgler, T. Schaub, R. Weisendanger, and H.-J. Günterodt, Surf. Sci. **285**, 1 (1993).
[6] F.C. Frank, Disc. Faraday Soc. **5**, 48 (1949).
[7] T.E. Mitchell, W.T. Donlon, K.P.D. Lagerlof, and A.H. Hauer in *Structure of Dislocations in Oxides,* edited by R.E. Tressler and R.C. Brandt (Plenum Press, New York, 1984), pp. 125-139.
[8] J.P. Hirth and G.M. Pound, J. Chem. Phys. **26**, 1216 (1957).
[9] S.V. Yanina and C.B. Carter, in preparation.

Mat. Res. Soc. Symp. Vol. 620 © 2000 Materials Research Society

Resolving the Control of Magnesium on Calcite Growth: Thermodynamic and Kinetic Consequences of Impurity Incorporation for Biomineral Formation

Kevin J. Davis, Patricia M. Dove and James J. De Yoreo*
School of Earth and Atmospheric Sciences,
Georgia Institute of Technology,
Atlanta, GA 30332
*Department of Chemistry and Materials Science,
Lawrence Livermore National Laboratory,
Livermore, CA 94550

ABSTRACT

Magnesium is a key determinant in $CaCO_3$ biomineral formation and has recently emerged as an important paleotemperature proxy. Atomic force microscopy (AFM) was used to determine the fundamental thermodynamic and kinetic controls of Mg^{2+} on calcite morphology and growth. Comparison of directly measured monomolecular step velocities ($v_{s\pm}$) to theoretical crystal growth impurity models demonstrated calcite inhibition due to enhanced mineral solubility through Mg^{2+} incorporation. Terrace width (λ) measurements independently supported an incorporation mechanism by indicating a shift in the effective supersaturation (σ_{eff}) of the growth solutions in the presence of Mg^{2+}. This study resolves the controversy over the molecular-scale mechanism of calcite inhibition by Mg^{2+} and provides an unambiguous model for the thermodynamic and kinetic consequences of impurity incorporation into $CaCO_3$ biominerals.

INTRODUCTION

The complex processes of biomineralization offer exciting prospects for the development of novel materials based upon biomimetic strategies [1-3]. In addition, $CaCO_3$ biomineral formation has widespread implications for global biogeochemical cycles. Biogenic carbonate sediments are recognized as biogeochemically-significant due to their role in mediating ocean chemistry and atmospheric CO_2 concentrations. Further, the trace element composition of $CaCO_3$ biominerals has been shown to reflect the chemical and physical environments in which they formed, resulting in their use as important tools in paleoclimate determination [4].

Biomineralization occurs in the complex aqueous systems that are characteristic of natural environments. A critical step in understanding biomineral formation is to determine the fundamental interactions of common inorganic aqueous species with the growing mineral surface. Due to its ubiquity in natural waters, Mg^{2+} is the principal modifier of $CaCO_3$ morphology and growth in biogeochemical environments [5]. More importantly, the presence of magnesium in calcium carbonate biominerals has been identified as an invaluable paleothermometer that is less susceptible to changes in salinity and polar ice volume than other proxies [6-9]. However, despite the contemporary need to understand paleoclimates, the physical basis by which Mg^{2+} modifies carbonate growth has yet to be discerned. In fact, the molecular-scale mechanism by which Mg^{2+} inhibits calcite continues to be the source of considerable controversy in the geochemical community. This disagreement is driven by conclusions from bulk studies that attribute calcite growth inhibition to either *step-blocking* by Mg^{2+} adsorption and slow-dehydration [10-13], or to increases in mineral solubility associated with Mg^{2+} *incorporation* [14]. Since calcite mineralization is fundamentally a crystal growth phenomenon, a mechanistic understanding of calcite growth inhibition by Mg^{2+} may best be achieved in the context of crystal growth theory. This study resolves the role of Mg^{2+} in $CaCO_3$ biomineral formation through molecular-scale determination of the fundamental thermodynamic and kinetic controls on calcite growth by comparing *in situ* experimental measurements to theoretical crystal growth impurity models.

EXPERIMENTAL DETAILS

Fluid-cell atomic force microscopy (AFM) was used to make *in situ* observations of calcite crystallization onto a seed crystal in a flow-through environment [15]. Single-sourced growth spirals emanating from screw dislocations on the calcite surface were imaged in Contact Mode (Digital Instruments, Santa Barbara) under precisely controlled solution conditions at 25°C. Supersaturated growth solutions were carefully prepared from $NaHCO_3$ (Aldrich, 99.7+% A.C.S reagent), $CaCl_2 \cdot H_2O$ (Aldrich, 98+%, A.C.S. reagent) and $MgCl_2$ (Aldrich, 99.99% ReagentPlus™) using deionized water (resistance > 19MΩ). ICP-MS/AES was used to ensure the purity of all the reactants used in this study. The pH of each growth solution was adjusted to 8.50, the ionic strength was fixed between 0.115-0.119, and the ratio of calcium to carbonate activity held in the range 0.99-1.01. The P_{CO2} was fixed by using a closed-system configuration. Finally, the chemical speciation of each solution was rigorously modeled using a numerical code that implemented the Davies equation for activity determinations [16]. Monomolecular step velocities ($v_{s\pm}$) and terrace widths (λ) were directly measured on growth spirals as a function of both supersaturation (σ) and magnesium activity. The rate of solution input was adjusted to yield step velocities that were independent of flow-rate, thereby ensuring that growth was not limited by mass transport to the surface. Step velocity was determined as displacement from a fixed reference point (*i.e.* the dislocation source) or by changes in the apparent step orientation [17]. Terrace widths were simply determined by measuring the distance between adjacent steps.

The supersaturation, σ, is defined by

$$\sigma \equiv \frac{\Delta\mu}{k_b T} = \ln\left(\frac{a}{a_e}\right) \quad (1a)$$

or for $CaCO_3$ in aqueous solution

$$\sigma = \ln\left(\frac{a_{Ca^{2+}} a_{CO_3^{2-}}}{K_{sp}}\right) \quad (1b)$$

where $\Delta\mu$ is the change in chemical potential per molecule, k_b is the Boltzmann constant, T is absolute temperature, a and a_e are actual and equilibrium activity products, K_{sp} is the equilibrium solubility at the ionic strength of the experimental solutions, and $a(i)$ is the activity of the ith species. K_{sp} ($10^{-8.48}$) was calculated from the activities at which measured step speeds went to zero. Here we present *in situ* thermodynamic and kinetic measurements for σ equal to 0.98, 1.20, and 1.40 at various magnesium impurity concentrations. Each data point represents a separate experiment with a unique solution composition.

RESULTS AND DISCUSSION

Calcite spirals grown in the pure system generated well-formed polygonized hillocks with straight step-edges (**figure 1a**). The addition of Mg^{2+} to the growth solutions modified the kink site activity of calcite, resulting in significant step-edge roughening (**figure 1b**). Figure 1b further demonstrates that Mg^{2+} preferentially interacts with steps along the negative direction ($[\overline{4}41]$ and $[48\overline{1}]$). This result is a direct consequence of the nonequivalent kink site structure of the (+) and (-) step-edge directions and is indicative of the importance of kink site structure in determining crystal growth behavior. Larger additions of Mg^{2+} significantly altered hillock morphology by causing a progressive rounding of the +/- corners of the spiral. A mechanistic understanding of these morphological observations and the physical basis by which Mg^{2+} interacts with the calcite surface requires the measurement of fundamental crystal growth parameters, such as step velocity ($v_{s\pm}$) and terrace width (λ).

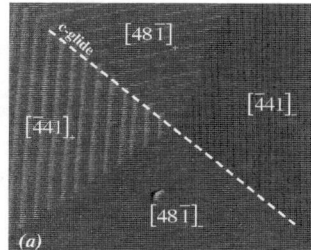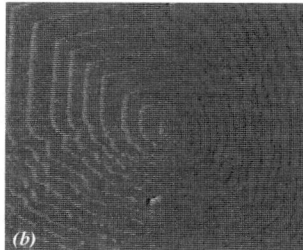

Figure 1 *Calcite spiral growth hillock with step directions and c-glide symmetry labeled, grown in (a) pure solution (σ=0.98) and (b) Mg^{2+}-bearing solution (σ=0.98; a_{Mg2+}=3.21x10^{-5}).*

Monomolecular step velocities were measured to significantly decrease as a function of magnesium concentration for steps along both the (+) and (-) directions (**figure 2**). This reduction in step migration rates was linearly dependent upon Mg^{2+} concentration and greater amounts of magnesium were required to halt growth with increasing supersaturation. The comparison of these kinetic observations with crystal growth impurity models (**figure 3**) clearly support an incorporation mechanism for the inhibition of calcite by Mg^{2+}. As the magnesium concentration increases in the growth solutions, larger amounts of Mg^{2+} enter the calcite lattice, enhancing the solubility of the growing crystal. This causes the growth solutions to exhibit a lower effective supersaturation (σ_{eff}) as the equilibrium point of the system is shifted toward higher activities. The shift in equilibrium activity accounts for the region of positive supersaturation where no growth occurs and the slower growth rate observed in the presence of Mg^{2+} [18,19].

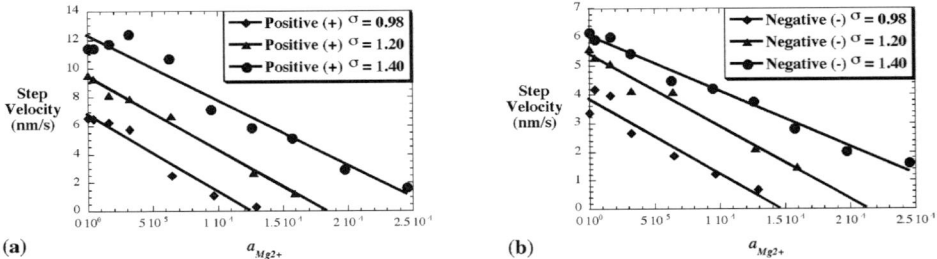

(a) a_{Mg2+}

(b) a_{Mg2+}

Figure 2 Step velocity versus magnesium activity at multiple supersaturations for **(a)** *steps along the positive (+) directions and* **(b)** *steps along the negative (-) directions.*

(a) *Impurity Concentration*

(b) *Impurity Concentration*

Figure 3 Models for impurity inhibition of crystal growth. **(a)** *Incorporation mechanism inhibition: step velocity decreases linearly with increasing impurity concentration. Higher supersaturations (where $\sigma_3 > \sigma_2 > \sigma_1$) result in a faster growth rate at the same impurity concentration.* **(b)** *Step blocking mechanism of inhibition: step velocity is unaffected by the impurities present in solution until a threshold impurity concentration is reached, which results in a sudden decrease in growth rate. Higher supersaturations (where $\sigma_3 > \sigma_2 > \sigma_1$) require a greater impurity concentration for a cessation of growth to occur.*

In parallel with these kinetic observations, measurements of terrace width provided an independent test of whether Mg^{2+} shifts the apparent thermodynamic properties of the system. In the absence of impurities, terrace widths increase with increasing supersaturation (σ) according to the following expression of the classic Gibbs-Thomson relation:

$$\lambda = \frac{2.04\Gamma\omega\alpha}{k_B T\sigma} \tag{2}$$

where λ is the terrace width (nm), Γ is a factor that takes into account the dependence of step speed on step length, ω is the specific molecular volume (6.13×10^{-23} cm^3/molecule for calcite), α is the step-edge free energy per unit step height, k_B is the Boltzmann's constant, T is the temperature (K), σ is the previously defined supersaturation, and the quantity 2.04 is a factor related to the geometry of the calcite hillock [15]. A general result from equation 2 is that a decrease in the σ_{eff} of the growth solutions would correspond to a linear increase in the observed terrace width (λ). Figure 4 shows that terrace width was indeed found to increase linearly with increasing magnesium activity. These thermodynamic effects on terrace width are summarized visually by the images in figure 5 which depict changes in monomolecular step density as a function of supersaturation and magnesium concentration. The reduction in σ_{eff} implied by the measured increase in terrace width is further evidence of enhanced calcite solubility through Mg^{2+} incorporation. Therefore both the thermodynamic and kinetic measurements of this study independently support an incorporation mechanism for the inhibition of calcite by Mg^{2+}.

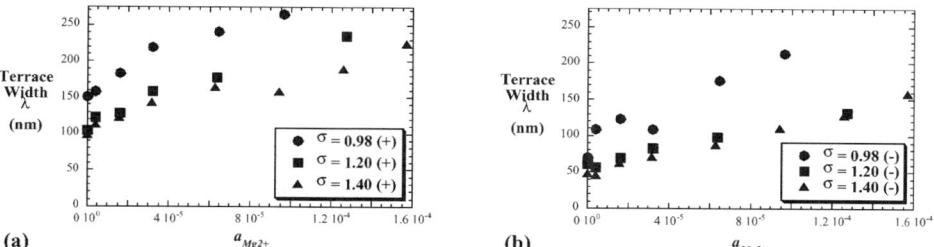

(a)   **(b)**

Figure 4 *Terrace width versus magnesium activity at multiple supersaturations for* ***(a)*** *steps along the positive (+) directions and* ***(b)*** *steps along the negative (-) directions.*

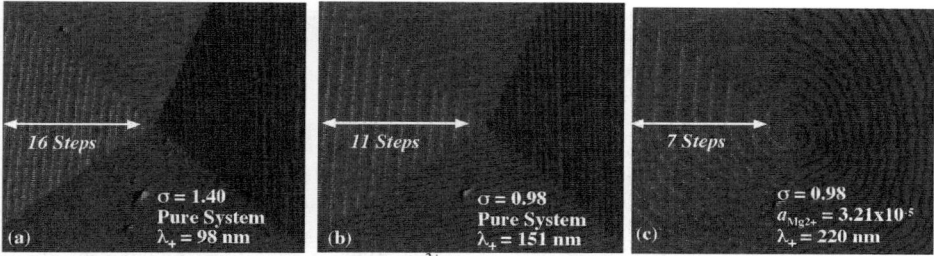

Figure 5 *Effect of supersaturation and Mg^{2+} on terrace width. Decreasing supersaturation and increasing a_{Mg2+} induces an increase in the observed terrace width, resulting in lower step density. The presence of Mg^{2+} in the growth solutions enhances the solubility of the mineral resulting in a terrace width increase similar to that expected from a reduction in supersaturation (σ). (All images $3\mu m$ x $3.5\mu m$)*

CONCLUSIONS

Observations of hillock morphology clearly demonstrated that Mg^{2+} modifies the kink site activity of step-edges on the calcite surface. Monomolecular step velocity ($v_{s\pm}$) measurements matched the theoretical crystal growth impurity model for growth inhibition by incorporation. Measurements of terrace width (λ) independently supported this finding by indicating a decrease in the effective supersaturation (σ_{eff}) of the growth solutions with increasing magnesium concentration. This study resolves the controversy over the molecular-scale mechanism by which magnesium inhibits calcite growth and defines the physical role of Mg^{2+} in mediating $CaCO_3$ morphology and growth. Finally, this investigation provides a model system for the thermodynamic and kinetic consequences of impurity incorporation into biominerals by unambiguously showing that Mg^{2+} inhibits calcite growth through modification of thermodynamic properties induced by incorporation.

ACKNOWLEDGEMENTS

This work was supported by the Geosciences Research Program, Basic Energy Sciences, U.S. Department of Energy through grant number DE-FG05-95-ER14517 and was performed under the auspices of Lawrence Livermore National Laboratory under contract W-7405-Eng-48.

REFERENCES

1. Mann, S. (1995) Biomineralization, the Inorganic-Organic Interface, and Crystal Engineering. In *Biomimetics: Design and Processing of Materials*. Edited by M. Sarikaya and I.A. Aksay. American Institute of Physics, New York, pp. 91-116.
2. Mann, S. (1996) Biomineralization and Biomimetic Materials Chemistry. In *Biomimetic Materials Chemistry*. Edited by S. Mann. VCH Publishers, Inc., New York., pp. 1-37.
3. S. Mann, B. R. Heywood, J. M. Didymus, S. Rajam, V.J. Wade and J.B.A. Walker (1990) Biomineralization: New Routes to Crystal Engineering. In *Materials Synthesis Utilizing Biological Processes*. Edited by P.C. Rieke, P.D. Calvert and M. Alper. Materials Research Society Symposium Proceedings, Volume 174.
4. Lea, D. W. (1999) Trace elements in foraminiferal calcite. In *Modern Foraminifera*. Edited by B.K. Sen Gupta. Chapman & Hall, London.
5. Folk, R.L. (1974) The Natural History of Crystalline Calcium Carbonate: Effect of Magnesium Content and Salinity. *Journal of Sedimentary Petrology* **44**, 40-53.
6. Lear, C.H., Elderfield, H. and Wilson, P.A. (2000) Cenozoic Deep-Sea Temperatures and Global Ice Volumes from Mg/Ca in Benthic Foraminiferal Calcite. *Science* **287**, 269-272.
7. Purton, L.M.A., Shields, G.A., Brasier, M.D. and Grime, G.W. (1999) Metabolism controls Sr/Ca ratios in fossil aragonitic mollusks. *Geology* **27**, 1083-1086.
8. Mitsuguchi, T., Matsumoto, E., Abe, O., Uchida, T. and Isdale, P.J. (1996) Mg/Ca Thermometry in Coral Skeletons. *Science* **274**, 961-963.
9. Dwyer, G.S., Cronin, T.M., Baker, P.A., Raymo, M.E., Buza, J.S. and Corrége, T. (1995) North Atlantic Deepwater Temperature Change During Late Pliocene and Late Quaternary Climatic Cycles. *Science* **270**, 1347-1351.
10. Reddy, M.M. and Wang, K.K. (1980) Crystallization of Calcium Carbonate in the Presence of Metal Ions. I. Inhibition by magnesium ion at pH 8.8 and 25°C. *Journal of Crystal Growth* **50**, 470-480.
11. Mucci, A. and Morse, J.W. (1983) The incorporation of Mg^{2+} and Sr^{2+} into calcite overgrowths: influences of growth rate and soluiton composition. *Geochimica et Cosmochimica Acta* **47**, 217-233.
12. Reddy, M.M. (1986) Effect of Magnesium Ions on Calcium Carbonate Nucleation and Crystal Growth in Dilute Aqueous Solutions at 25°C. In *U.S. Geological Survey Bulletin 1578: Studies in Diagenesis*. Edited by F.A. Mumpton.

13. Gutjahr, A., Dabringhaus, H. and Lacmann, R. (1996) Studies of the growth and dissolution kinetics of the $CaCO_3$ polymorphs calcite and aragonite: II. The influence of divalent cation additives on the growth and dissolution rates. *Journal of Crystal Growth* **158**, 310-315.
14. Berner, R.A. (1975) The role of magnesium in the crystal growth of calcite and aragonite from sea water. *Geochimica et Cosmochimica Acta* **39**, 489-504.
15. Teng H. H., Dove P. M., Orme C. A., and De Yoreo, J. J. (1998) The thermodynamics of calcite growth: A baseline for understanding biomineral formation. *Science* **282**, 724-727.
16. Papelis C, Hayes K. F., and Leckie J. O (1988) A program for the computation of chemical equilibrium composition of aqueous batch systems including surface-complexation modeling of ion adsorption at the oxide/solution interface. Dept. of Civil Engi., *Technical Report 306*, Stanford University.
17. Land, T. A., De Yoreo, J. J. and Lee, J. D. (1997) An *in situ* AFM investigation of canavalin crystallization kinetics. *Surf. Sci.* **384**, 136-155.
18. Voronkov, V. V. and Rashkovich, L. N. (1992) Influence of a mobile adsorbed impurity on the motion of steps. *Sov. Phys. Crystallogr.* **37(3)**, 289-295.
19. van Enckevort, W.J.P. and van den Berg, A.C.J.F. (1998) Impurity blocking of crystal growth: a Monte Carlo study. *Journal of Crystal Growth* **183**, 441-455.

Mat. Res. Soc. Symp. Vol. 620 © 2000 Materials Research Society

Novel Studies of Roughening of the Prism Plane of Ice

Ann-Marie Williamson & Alex Lips
Unilever Research Colworth, Colworth House, Sharnbrook, Bedford, MK44 1LQ, UK

ABSTRACT

A novel technique for examining kinetic roughening of crystals is described, and applied to the study of the prism plane of ice in contact with aqueous fructose solution. The technique can be generally applied to crystals that roughen at low driving forces. Since the residual driving force for growth utilised is that due to Ostwald ripening, this technique also facilitates simultaneous quantification of ensemble growth kinetics and crystal anisotropy during ripening. The driving force required for kinetic roughening, and the step, or ledge, free energy of this plane of ice show an approximately linear variation with temperature over the experimental temperature range $-13°C$ to $-17°C$. Whilst we can conclude that the thermodynamic roughening temperature (T_R) is higher than $-13°C$, its precise value, and conformance or otherwise of the roughening transition with Kosterlitz Thouless scaling, can not be concluded from the current data set.

INTRODUCTION

The state of roughness of a crystal surface affects microstructural evolution (e.g. the size, shape and degree of connectedness of crystals) for crystal/fluid systems. This paper describes a new technique developed to determine the kinetic roughening transition for crystal interfaces that kinetically roughen at low driving forces, and involves a detailed investigation of isothermal coarsening of a dilute dispersion of crystals, with quantification of their growth morphologies during coarsening. Furthermore, since the residual driving force for growth of the crystals results solely from Ostwald ripening, this technique has the advantage that, in addition to mapping the kinetic roughening transition, the kinetics of ensemble ripening, the crystal size distribution, and the state of crystal anisotropy are simultaneously monitored. The crystal interface studied is that between the prism plane of ice and an aqueous fructose solution; fructose providing a convenient solute for controlling the dissolution temperature of ice in the temperature range of interest. This has been identified from recent work by Maruyama et al.[1,2], based on the use of pressure to vary the freezing point of ice, that suggests that the thermodynamic roughening temperature (T_R) of this plane of ice in contact with water lies in the range $-13°C$ to $-17°C$.

The technique looks for the onset of faceting of the prism plane of ice by following changes in the morphology of crystals oriented with their basal planes normal to the observation direction, i.e. crystals which start with a high degree of circularity in their 2-D profile, but develop hexagonal symmetry as the prism-plane kinetic faceting transition is traversed. Fourier analysis is used to quantify changes to the symmetry of the crystals. These changes are correlated with the measured supersaturation of the ensemble in its asymptotic ripening limit, and estimates are made of the magnitude and temperature dependence of the step free energy of the prism plane of ice in this temperature range.

EXPERIMENTAL DETAILS

The experiment was performed using an automated cryo-microscope described in detail elsewhere[3]. The microscope stage was developed to maintain temperature to a precision of better than +/-10^{-3}°C; the temperature of the stage was calibrated using the melting points of decane (-29.7°C) and ice (0.0°C).

Table I Initial concentration of fructose used, concentration of fructose in coexisting liquid phase, and calculated values of F for each isothermal experiment

Temp /°C	Initial mass percentage of fructose in aqueous solution	Mass percentage of fructose in liquid phase at temperature employed	$F = (r_{mean,growing}/r_{mean,ensemble})$
-13.5	46%	52.1%	1.188 ± 0.017
-14.5	48%	53.4%	1.130 ± 0.013
-15.5	50%	54.7%	1.127 ± 0.018
-16.5	52%	55.8%	1.116 ± 0.014
-17.5	54%	56.9%	1.128 ± 0.008

For each experiment a 0.6 µl sample of aqueous fructose (Fisons AR grade) solution was placed in the sample crucible. The initial sample concentration used for each isothermal holding temperature is given in Table I. The sample thickness was fixed at 30 µm by use of a stainless steel spacer, and the sample sealed from the atmosphere by placing a cover glass on top of this spacer. The crucible was retained in position by the sample carrier, and moved so that it lay on a portion of the stage thermally bridged to ambient. Dry nitrogen was repeatedly flushed through the stage (to displace air) before sealing it, so as to minimise condensation. The temperature controlled portion of the stage was precooled to -100°C (a silver cover placed over this portion of the stage minimises temperature gradients normally through the sample), and the sample rapidly translated onto its surface so as to ensure its rapid vitrification. The sample was then left to reach thermal equilibrium for 2 minutes before raising its temperature, at a rate of 5°C/min, to the isothermal holding temperature of the experiment. This protocol results in a nucleation dominated ensemble of small ice crystals. Time zero for the experiment is defined as the time when the sample first reaches the isothermal holding temperature. A raster pattern was pre-set for the motion of the sample carrier during the experiment, so that a number of selected spatial locations within the sample were regularly revisited. Images were acquired sequentially at each location throughout the experiment; the ice crystals were allowed to ripen until no obvious further change in morphology was detected. The duration of the experiments was typically 8 – 24 hours.

For each experiment, images at one spatial location and microscope magnification were selected for analysis, so that they reflected the change in crystal morphology expected upon traversing the kinetic faceting transition i.e. initial images contained rounded ice crystals, later images were of partially faceted crystals (see Figure 1). Each image was then segmented to obtain a binary image, and shape criteria used to select crystals for analysis which were oriented with their basal plane normal to the viewing direction. The aim of the procedure was to analyse as many crystals as possible, rejecting only those crystals which were either accreted, too small to provide reliable data when subjected to harmonic analysis[3], or aligned in such a way as to render the results of harmonic analysis meaningless. Analysing only crystals which had an area > 300 pixels, an aspect ratio[†] greater than typically 0.85, and a form factor[‡] greater than typically 0.80 satisfied this aim. The selected crystals in each image were then subjected to harmonic analysis, and their area determined.

The resulting basis set of data for each image was necessarily noisy, since each ice crystal analysed was subject to a local curvature excess chemical potential dependent on both its in view, and out of view, dimensions. The noise in the data was reduced by first rejecting data from crystals having a radius smaller than the average of the ensemble (the assumption being that the latter corresponds approximately to the critical radius of the ensemble, and that crystals smaller than this would be dissolving and therefore rough[4,5,6]), and then averaging both the normalised sixth harmonic, A_6/A_0, and radius of the resulting crystals. The following parameters were determined: the mean radius of the ensemble ($r_{mean,ensemble}$), the mean radius of the sub-set of crystals assumed to be growing ($r_{mean,growing}$), the mean normalised sixth harmonic for this subset, and the number of crystals averaged.

Examination of crystals oriented with their prism plane normal to the direction of observation enabled the anisotropy of the crystals to be characterised. For crystals oriented in this fashion we measured the minimum feret, the feret orthogonal to the minimum, and identified the plane corresponding to the minimum feret. Time-resolved studies of the dependence of the ratio of extent of expressed crystal planes, with measured extent of basal plane, were undertaken.

DISCUSSION

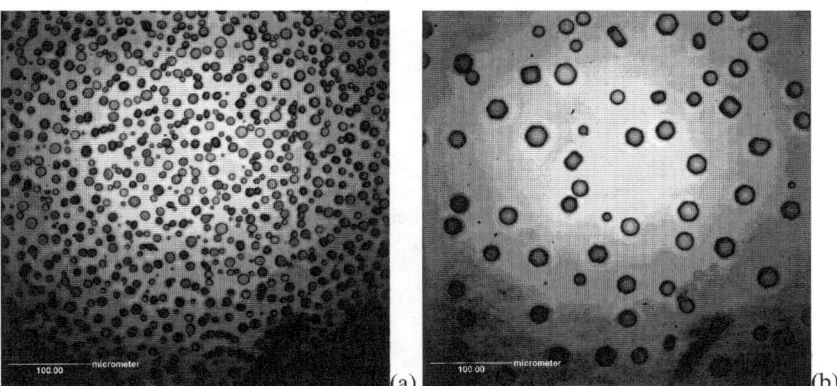

Figure 1 Experimental images obtained on ripening at –15.5°C for (a) 1875 s and (b) 39438 s.

[†] Defined as the ratio of the minimum to maximum object feret
[‡] Measurement of circularity of a shape. Formfactor = (4 π Area) / Perimeter[2]

The results of the procedure described above are given in Figure 2, which shows the average change in A_6/A_0 with mean radius of the subset of crystals considered to be growing at each temperature. It is apparent that we can clearly delineate a roughening to faceting transition as the supersaturation diminishes. Each curve represents an isothermal transition, both the amplitude and mid-point of which show the expected qualitative dependence on temperature; crystals facet at smaller radii at lower temperatures. The data is well fitted by a sigmoidal function of the form:

$$\frac{A_6}{A_0} = \left(\frac{a_r - a_f}{1 + e^{(r-r_0)/dr}} \right) + a_f \qquad (1)$$

with r_0 the radius of growing crystals corresponding to the centre of the transition, dr the width of the transition, a_r the initial value of A_6/A_0 when the crystals are still rough, and a_f the final value of A_6/A_0 attained by the faceted crystals. We expect the latter value to be close to that expected for crystals in the Wulff condition.

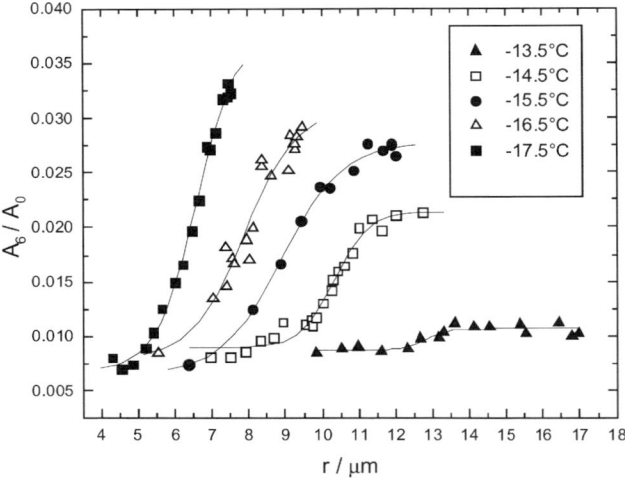

Figure 2 Variation of the mean normalised sixth harmonic (A_6/A_0) of growing crystals viewed normally to their basal plane, with their mean radius (r) for each of the experimental temperatures indicated in the legend.

In general, the supersaturation of an isothermal ensemble of spherical crystals undergoing late-stage coarsening is given by[4]:

$$\frac{\Delta \mu}{k_B T} = \frac{2\sigma v_{mol}}{r_c k_B T} \qquad (2)$$

with r_c the "critical" radius of the whole ensemble, σ the inter-phase surface energy, v_{mol} the molecular volume of water, k_B the Boltzmann constant and T the temperature. Crystals of critical radius are in equilibrium with the solution, so their Gibbs Thompson curvature chemical potential then balances the supersaturation and is given by equation (2). Since the ratio of mean radius of the ensemble to r_c is constant in the asymptotic limit, with the value of this constant (k_1) close to 1[4,5,6] (equal to 1 for diffusion limited ripening and 8/9 for interface kinetics limited ripening), we estimate the curvature chemical potential of the critical radius (supersaturation) at transition midpoints $r = r_0$ from measurement of the mean radius of the ensemble, and on the basis of a value of σ/k_1 of 0.03 Nm^{-1}. Equation (2) assumes that the crystals are spherical. It is possible to approximately correct for crystal anisotropy by recognising that the excess chemical potential of the prism plane due to finite size is given by[3]:

$$\frac{\Delta\mu_{PRISM}}{k_B T} = \left\{ \frac{4v_{mol}\sigma_{PRISM}}{\sqrt{3}\,xk_B T} \right\} \left[0.5\left(1 + \frac{A}{A_{EQU}}\right) \right] \qquad (3)$$

where $x\sqrt{3}$ is the shortest extent of the basal plane, with x the length of one side of a hexagon, A is the ratio of plane lengths ($x\sqrt{3}/y$) where y is the length of the prism plane, and $A_{EQU} = (\sigma_{PRISM}/\sigma_{BASAL})$. In order to apply the correction, an estimate must be made of the Wulff shape. Assuming weak temperature dependence for the surface energies of both expressed crystal planes, and therefore a constant Wulff crystal ratio of 1.4 (obtained from a previous study of ripening at -19°C[3]) we correct for anisotropy from a determination of the measured plane length ratio at the transition i.e. when $r = r_0$. Figure 3 shows for transition mid-points the temperature dependence of our estimate of supersaturation obtained by these procedures. The scaling behaviour of the transition is found to be substantially unaffected by the correction for crystal anisotropy.

The local driving force for growth experienced by a crystal within an ensemble is the difference between the supersaturation and the curvature chemical potential of that crystal. Averaged local driving forces can be expected to be *proportional* to the supersaturation of the ensemble provided late stage coarsening applies with size distributions in their *asymptotic* limit[4,6]. Our best estimate of the proportionality constant is obtained from the experimentally determined value of (F-1)/F (F is defined as the ratio of r mean,growing to r mean,ensemble and is given in Table I). The step, or ledge, free energy can be obtained from the calculated driving force required to kinetically facet the prism plane of ice if we make the additional assumptions that:
(i) Ice plane surface energies are significantly less temperature dependent than the corresponding step free energies over the temperature range considered.
(ii) The asymptotic distribution is similarly temperature insensitive. That this condition is largely met in the present study is supported by the invariance of the factor F (see Table I).
Figure 4 shows the temperature dependent step free energy, $\gamma/k_B T$, calculated upon making these assumptions. The calculated value is dependent on the classical nucleation criteria used[7]:
Criterion N1 - A surface will grow kinetically roughened if the Gibbs energy needed for the formation of a 2D nucleus (ΔG^*) is less than approximately $k_B T$.
Criterion N2 – A surface will grow kinetically roughened if the radius of a critical 2D nucleus r^* is smaller than approximately half a growth unit.

The step free energy is found to vary approximately linearly with temperature over the range considered suggesting that, if the system conforms to Kosterlitz Thouless (KT) scaling [8,9,10,11,12], T_R is likely to be above -8°C[13]. The discrepancy between this value and that due to Maruyama et al.[1,2] probably results from the non-attainment of true equilibrium openly acknowledged in the latter study. However, it is possible that roughening in this system is a higher order transition than implied by KT scaling; it should also be noted that the current study uses fructose to control the dissolution temperature of ice, and that this solute may influence its roughening behaviour.

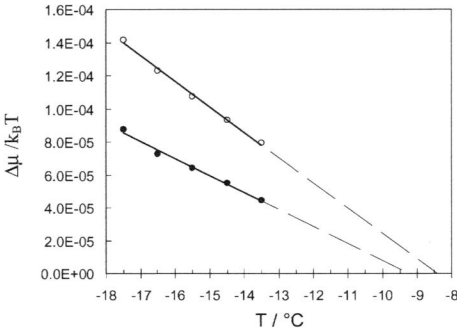
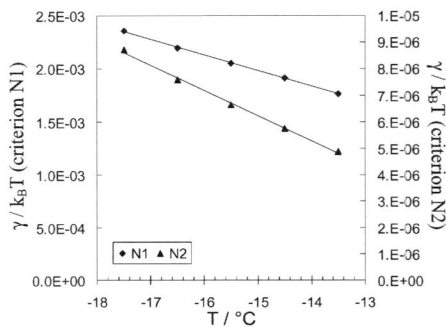

Figure 3 Temperature dependence of the supersaturation of the ensemble ($\Delta\mu/k_BT$) at the mid-point of the kinetic faceting transition: (i) assuming spherical crystals (•) (ii) correcting for crystal anisotropy (o).

Figure 4 Temperature dependence of the step free energy (γ/k_BT) determined using classical nucleation criteria N1 and N2 as indicated in the legend[7]

ACKNOWLEDGEMENTS

The authors would like to thanks Profs. Allan Clark and Denver Hall for discussions and critical review of this work.

REFERENCES

1. M. Maruyama, T. Nishida, T. Sawada, *J. Phys. Chem. B*, **101**, 6151 (1997).
2. M. Maruyama, Y. Kishimoto, T. Sawada, *J. Crystal Growth*, **172**, 521 (1997).
3. A.-M. Williamson, A. Lips, A. Clark, D.G. Hall, *Faraday Discussions*, **112**, 31 (1999).
4. I.M. Lifshitz, V.V. Slyozov, *J. Phys. Chem. Solids*, **19**, 35 (1961).
5. A. Bhakta, E. Ruckenstein, *J. Chem. Phys.*, **103**, 7120 (1995).
6. J.A. Marqusee, J. Ross, *J. Chem. Phys.*, **79(1)**, 373 (1983).
7. E. van Veenendaal, P.J.C.M. van Hoof, J. van Suchtelen, W.J.P. van Enckevort, P. Bennema, *Surf. Sci.*, **417**, 121 (1998).
8. F. Gallet, P. Nozières, S. Balibar, E. Rolley, *Europhys. Lett.*, **2**, 701 (1986).
9. L. Jörgenson, R. Harris, *Phys. Rev. E*, **47**, 3504 (1993).
10. X.-Y. Liu, P. van Hoof, P. Bennema, *Phys Rev. Lett*, **71**, 97 (1993).
11. J.M. Kosterlitz, D.J. Thouless, *J. Phys. C*, **6**, 1181 (1973).
12. M. Wortis, in *Chemistry and Physics of Solid Surfaces VII*, eds. R. Vanselow & R. Howe, Springer Verlag, Berlin (1988).
13. A.-M. Williamson, A. Lips, Unpublished work.

Mat. Res. Soc. Symp. Vol. 620 © 2000 Materials Research Society

Intrasectoral Zoning of Proteins and Nucleotides in Simple Crystalline Hosts

Miki Kurimoto, Loyd D. Bastin, Daniel Fredrickson, Pamela N. Gustafson, Sei-Hum Jang, Werner Kaminsky, Scott Lovell, Christine A. Mitchell, Jean Chmielewski,[1] Bart Kahr
Department of Chemistry, Box 351700, University of Washington,
Seattle, WA 98195-1700, U.S.A.
[1]Department of Chemistry, 1393 Brown Laboratories, Purdue University,
West Lafayette, IN 47907-1393, U.S.A.

ABSTRACT

Oriented gases of biopolymers in simple, single crystal hosts might be used to measure anisotropic molecular properties of analytes that could not otherwise be crystallized. Here we show two types of crystals as examples of the single crystal matrix isolation of biopolymers: green fluorescent protein in α-lactose monohydrate as a model system for studying the kinetic stabilization of biopharmaceuticals, and adenosine phosphates in potassium dihydrogen phosphate, a first step in the matrix isolation of oligonucleotides. In each case, the hosts undergo compositional zoning – both intersectoral and intrasectoral – during growth from solution. Intrasectoral zoning is evident by the selective luminescence of adjacent vicinal slopes of growth active hillocks. Nucleotides furthermore distinguish between symmetry related growth sectors enantioselectively.

INTRODUCTION

Efforts to understand how and why large guest molecules enter simple carboxylic acid host crystals during growth from solution have taken on a new urgency in light of the explosive growth of matrix-assisted laser desorption ionization mass spectrometry (MALDI-MS), a revolutionary process in which biopolymer analytes precipitated with a crystalline acid matrix, are propelled into the gas phase upon irradiation [1]. The success of the MALDI experiment depends critically on the choice of host crystal but the characteristics of a good host such as 2,5-dihydroxybenzoic acid, or of a poor one such as the constitutional isomer 3,5-dihydroxybenzoic acid, are still not understood. The inventors of MALDI-MS demonstrated that when large crystals of matrices were grown in the presence of proteins, cleaved, and irradiated on the fresh surface, the signal in the mass spectrum was no less intense than that from the irradiation of a polycrystalline precipitate, suggesting that the growth of genuine mixed crystals is necessary for the transfer of energy from the laser to matrix to analyte [2]. This question has been studied intensively since 1991, but a current review of the subject is equivocal: "Protein incorporation into the crystals of solid MALDI matrices is helpful, but not a prerequisite [3]."

We demonstrated, however, that when single crystals of sinapic acid (3,5-dimethoxy-4-hydroxy-*trans*-cinnamic acid) grown in the presence of myoglobin were analyzed by MALDI-MS, the intensity of the protein signal in the mass spectrum was highly dependent upon the region of the crystal irradiated with the focused laser. This was a consequence of intersectoral zoning, the compositional partitioning of impurities in crystals between or among growth sectors not related to one another by symmetry. Irradiation on the largest uncolored (myoglobin free)

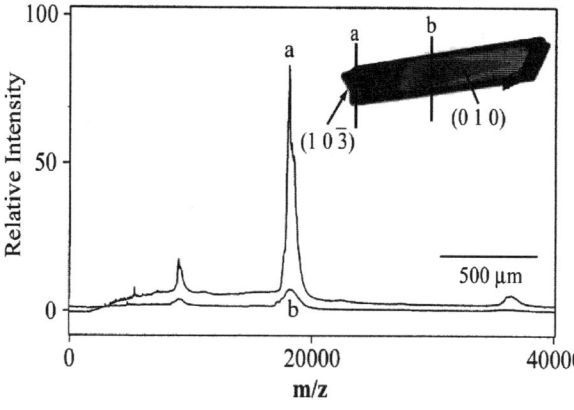

Figure 1. *MALDI mass spectrum of myoglobin in a sinapic acid single crystal.*

{010} growth sector produced a weak protein signal that increased 20-fold when the beam spilled over the edges of the crystal associated with the colored, myoglobin rich {10-3} growth sectors (Figure 1) [4,5]. Similar results were observed with other MALDI hosts and protein guests. During the course of these experiments we observed that certain crystals show remarkably general affinities for biopolymers. Accordingly, we embarked on further studies of the molecular recognition mechanisms that were operative during the single crystal matrix isolation of biopolymers. Two examples are discussed herein.

GREEN FLUORESCENT PROTEIN IN α-LACTOSE MONOHYDRATE

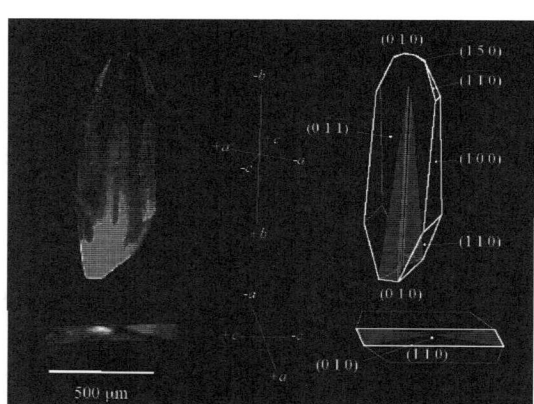

Figure 2. *Left: Fluorescence micrographs of LM/GFP. Right: Idealized representations. Micrographs were acquired by a CCD camera (Diagnostic Instruments, Inc.) connected to Olympus IMT-2 inverted microscope.*

Fluorescently labeled proteins have been adsorbed on and overgrown by crystals that compose biominerals [6,7]. We chose green fluorescent protein (GFP) [8] as a model protein dopant because it is known to fluoresce only in its native conformation. Upon denaturation, the interior of its β-barrel containing the luminophore is exposed, leading to fluorescence quenching. α-Lactose monohydrate (LM) was selected as a host because it is neutral and well formed (Figure 2) [9]. LM crystals are polar and grow unidirectionally along [010] [10]. GFP (Clontech, 1 mg/mL, 10 mM tris-HCl, pH 8, 10 mM EDTA) was added to saturated aqueous lactose solutions. After standing 3-4 days at room temperature, crystals 2 mm in length were obtained. The crystals contained one part GFP per 10^6 parts lactose.

When LM/GFP (notation indicates LM host crystal with GFP as guest) crystals were illuminated with a long wavelength UV lamp, they exhibited a bright green fluorescence (Figure 2) localized within a sharply defined pyramid corresponding to the (010) growth sector [11]. This intersectoral zoning indicated that the GFP was selected and overgrown by the (010) face in preference to the others. More importantly, the luminescence is strong evidence that GFP is in its native conformation inside the crystals. Measurement of the steady state fluorescence indicated that GFP was much more resistant to denaturation in the single crystal environment than in solution or in lyophilized lactose preparations [11]. LM/GFP crystals stored in our laboratories at room temperature for more than one year are still highly fluorescent.

Views of LM/GFP through the broad (0-1-1) face revealed a curious non-luminescent stripe running along the b axis. Analyses of the growth active (010) surface showed that this feature was the result of growth active hillocks [12]. Differential interference contrast microscopy of a pure LM crystal revealed a single polygonized hillock that partitioned the (010) surface into four vicinal slopes pair-wise related by two-fold symmetry [13]. Figure 3 compares an interference contrast image with atomic force micrographs made of the hillock center and of the topography of the steps on adjacent slopes. Similar images were previously made by Dincer *et al* [14].

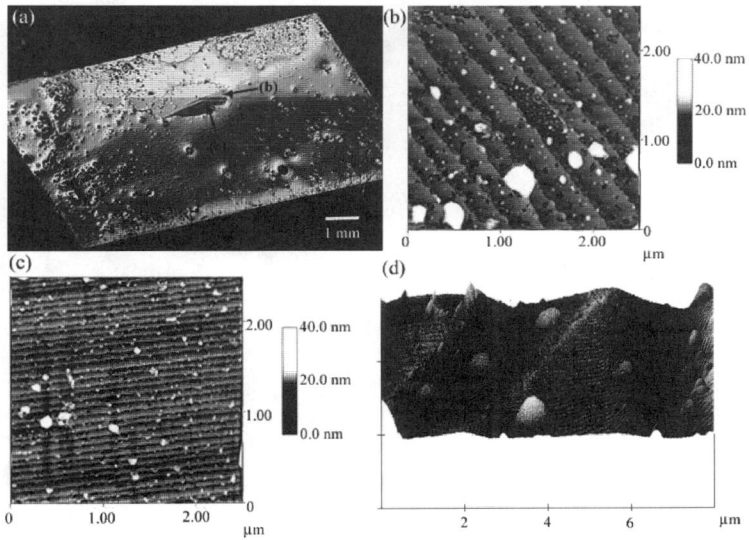

Figure 3. *(a) Differential interference contrast (DIC) micrograph of a pure LM crystal (010) face. Image is captured by a CCD camera attached to a Leica DMLM reflection microscope. (b,c) Atomic force microscopy (AFM) images of vicinal sectors where GFP would and would not be recognized, respectively. (d) AFM image of the hillock core; active slope is facing viewer. TappingMode AFM images (height mode)were obtained using Nanoscope IIIa (Digital Instrument), operated in air with a silicon probe. Scan rate = 1 Hz.*

Figure 4 compares the (010) DIC micrograph of LM/GFP with a fluorescence micrograph of the same surface indicating that only two of the four hillock slopes were fluorescent. This indicates that GFP recognized only the lateral slopes with greater step advancement velocity, an example of intrasectoral zoning. GFP added to those steps whose risers are populated by the faces of the sugar rings as opposed to the edges. In doped crystals, often hillocks are born and subsequently poisoned succumbing to new, active hillocks. By focusing through the (010) faces of lactose the sequence of active hillocks throughout the growth history are revealed by GFP luminescence (Figure 5).

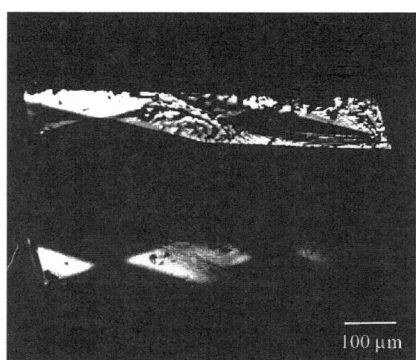

Figure 4. *Top: Differential interference contrast (DIC) micrograph of LM/GFPl (010) face. Bottom: Corresponding fluorescence micrograph.*

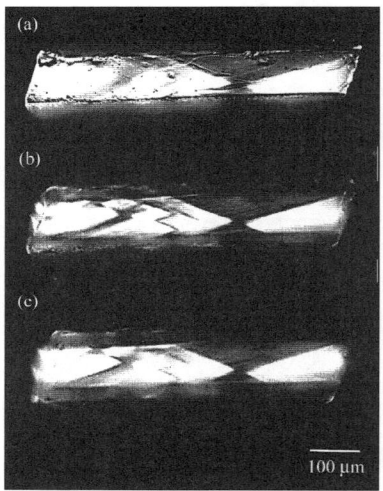

Figure 5. *Fluorescence micrographs of LM/GFP (010) face at (a) 0 μm, (b) 70 μm, and (c) 110 μm from the surface. These z-slices were generated using a Leica DMIRB incident-light inverted microscope.*

NUCLEOTIDES IN POTASSIUM DIHYDROGEN PHOSPHATE

The possibility of making oriented gases of DNA in crystals is appealing as a method for studying nucleic acid photophysics in the absence of conformational dynamics. As host for oligonucleotides, we chose KH_2PO_4 (KDP). Like DNA and RNA, KDP is a phosphate, and produces very large crystals whose growth mechanism has been extensively studied [15]. Moreover, we had previously shown that KDP is a remarkably general host for a wide variety of anionic organic dyes [16] and here extend our studies to adenosine triphosphate (ATP) and adenosine monophosphate (AMP).

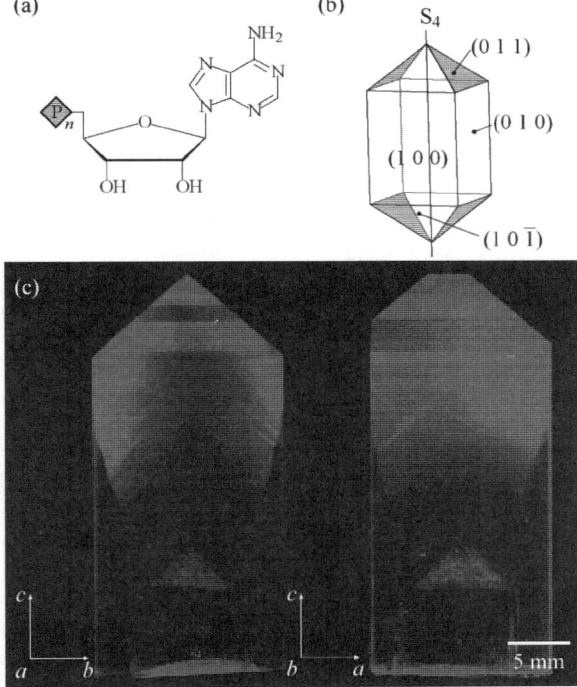

Figure 6. *(a) Adenosine phosphate. (b) Schematic diagram of KDP indicating the enantiomorphous (101) and (011) (shaded) faces. (c) Fluorescence photographs of KDP containing adenosine phosphates, viewed normal to the enantiomorphous a (left) and b (right) faces.*

A visible blue luminescence was localized in the pyramidal {101} growth sectors of KDP grown from a solution containing $(5 \times 10^{-4}$ M) ATP; the crystals contained one nucleotide per 5×10^4 KDP molecules. This is a common pattern of anion zoning in KDP [16]. Comparisons of the luminescence viewed through the *a* and *b* faces revealed that the (011) sector was considerably brighter than the (101) sector (Figure 6). As these sectors are mirror images of one another in the space group *I-42d*, the chiral nucleotides must have recognized these faces enantioselectively, as we have previously observed with hematein and KDP in the prismatic sectors [17,18]. Under the conditions of crystal growth ATP must undergo some hydrolysis. Therefore, the experiments were repeated with AMP because under the crystal growth conditions ATP undoubtedly underwent partial hydrolysis. Similar results were obtained.

The absolute configuration of the face that recognized KDP was determined by indexing the crystals and solving the structure using a Nonius KappaCCD diffractometer and its attendant software [19,20]. The indices refer to the coordinates as given by West [21]. Giving the ambiguity in assigning absolute configuration with automated indexing routines implemented in proprietary, commercial software packages, we checked our assignment by measuring the optical rotation. KDP crystals are achiral (point group D_{2d}) but nevertheless are optically active along the [100] and [010] directions with diad symmetry. In order to measure circular birefringence in directions that are also linearly birefringent, we employed the "tilter" method of Kaminsky and Glazer [22], a modification of the high-accuracy universal polarimetry (HAUP) method [23]. In this way, it was determined that the more highly fluorescent face of KDP doped with adenosine phosphates was (011) as correlated with the dextrorotatory [010] direction [24].

Zaitseva and coworkers have optimized KDP crystal growth conditions [25]. In their hands, amaranth (CI No. 16185) [26], an anionic dye that we had shown to have an exclusive affinity for the {101} surfaces of KDP, was both inter- and intrasectorally zoned [27]. This observation, manifest as a differential coloring of the three vicinal slopes of the hillocks on the pyramidal faces [28], required introduction of the dye during late growth thereby coloring only a thin surface layer so that patterns were not confounded by moving dislocation cores, seen, for example, in Figure 5. We have observed that AMP and ATP behaved similarly however, they selectively recognized C slopes, which were eschewed by amaranth (Figure 7).

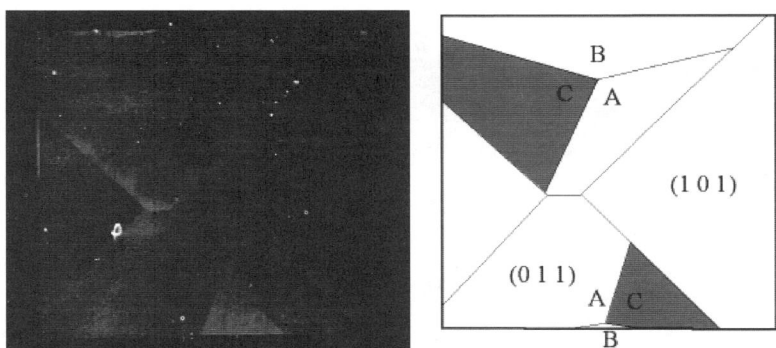

Figure 7. *Left: Fluorescence micrograph of a KDP/AMP. Right: Idealized representation illustrating that AMP recognizes the C slopes of growth hillocks on the (011) surfaces.*

DISCUSSION AND CONCLUSION

During the course of our research we have discovered a number of crystalline surfaces that have a remarkable ability to orient and overgrow guest molecules bearing neither size, shape, or constitutional similarity to the host crystal molecules and ions [29]. The continued generalization of single crystal matrix isolation requires an understanding of the characteristics that define a highly receptive face. Most receptive surfaces including the (011) face of KDP [30], the (110) and (021) faces of K_2SO_4 [31], and the (010) face of LM [30], are decorated with macroscopic hillocks that can be observed with a reflected visible light microscope equipped with a differential interference contrast prism. We then surmised that growth sectors easily doped with large guest molecules would be bounded by surfaces with macroscopic hillocks. We set out to test the idea with phthalic acid, an extraordinarily general host crystal.

Following the paths first trod by Lehmann [32], Gaubert [33], and Neuhaus [34], we found that a variety of dyes can color sectors of growing phthalic acid [35] and measurements of linear dichroism were used to reckon mixed crystal structure [36]. A particularly striking example is phthalic acid containing methyl red (CI No. 13020, one dye molecule per 2.6×10^3 phthalic acid molecules) in different states of protonation in the {010} and {021} growth sectors (Figure 8) yet showing comparable absorption anisotropies. Methyl red has a complex acid-base equilibrium due to the presence of a carboxylic acid group *ortho* to the azo bridge; protonated, deprotonated, neutral and zwitterionic forms have been spectroscopically differentiated in solution [37].

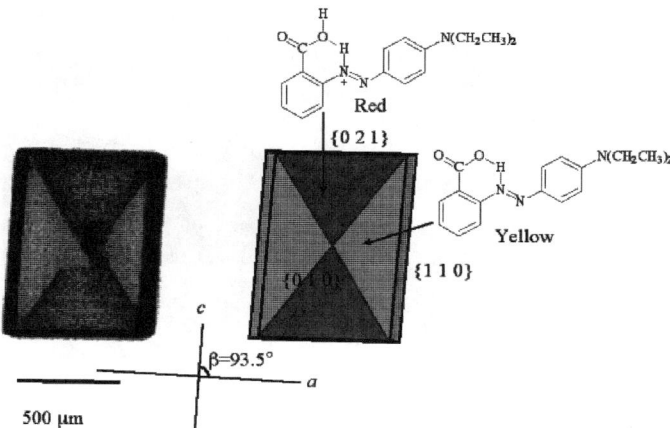

Figure 8. *Left: Phthalic acid/methyl red mixed crystal grown on a microscope slide. Right: Idealized representation with the corresponding methyl red protonation states.*

Phthalic acid crystals selectively incorporate the yellow, neutral form in the {010} sectors while {021} includes the red conjugate acid.

The growth mechanism of phthalic acid crystals has never before been described to the best of our knowledge. We observed that the {021} surfaces of pure crystals are optically flat; however, the addition of methyl red produced rounded, macroscopic hillocks (Figure 9).

What then is the relationship between macroscopic hillocks and mixed crystal growth? Macroscopic hillocks are typically formed through the process of step pinning by impurities and subsequent step bunching [38]. This process was shown explicitly with dye adsorbates by Mauri and Moret [39]. Accordingly, faces with macroscopic growth spirals are also likely to be those faces that have a tendency to strongly adsorb guests, either purposeful additives or adventitious impurities. To test this proposition, we grew crystals of potassium hydrogen tartrate, known to have well-formed hillocks on the *b* face [40], in the presence of a variety of dyes. In accord with Buckley, the dyes were indeed adsorbed and overgrown by (010) [41], and showed strong linear dichroism. The possible generalization of this supposition that macroscopic hillocks can serve to identify faces that are receptive to the incorporation of large guest molecules is currently subject to a variety of tests in our laboratories.

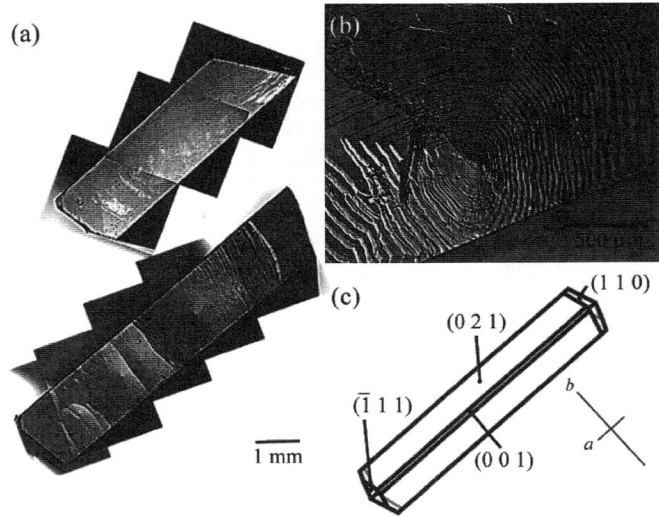

Figure 9. *(a) Composite DIC micrographs of phthalic acid crystals revealing (021) surfaces. Pure (top) and doped with methyl red (bottom). (b) A polygonized macrospiral formed on the phthalic acid/methyl red (021) surface. (c) Idealized representation.*

ACKNOWLEDGEMENTS

This work was supported by the National Science Foundation, the National Institutes of Health, the donors of the Petroleum Research Foundation of the American Chemical Society, and the University of Washington Center for Nanotechnology. We thank Richard W. Gurney for his generous assistance.

REFERENCES

1. F. Hillenkamp, M. Karas, R. C. Beavis, and B. T. Chait, *Anal. Chem.,* **63,** 1193A (1991).
2. K. Strupat, M. Karas, and F. Hillenkamp, *Int. J. Mass. Spec. Ion Proc.,* **111,** 89 (1991).
3. V. Horneffer, K. Dreisewerd, H.-C. Lüdemann, F. Hillenkamp, M. Lüge, and K. Strupat, *Int. J. Mass Spec.,* **185-187,** 859 (1999).
4. J. Chmielewski, J. L. Lewis, S. Lovell, R. Zutshi, P. Savickas, C. A. Mitchell, J. A. Subramony, and B. Kahr, *J. Am. Chem. Soc.,* **119,** 10565 (1997).
5. For experimental conditions see: S. J. Doktycz, P. J. Savickas, and D. A. Krueger, *Rapid Commun. Mass Spectrom.,* **5,** 145 (1991).

6. L. Addadi and S. Weiner, *Proc. Natl. Acad. Sci. USA,* **82,** 4110 (1985); *Mol. Cryst. Liq. Cryst.,* **134,** 305 (1986); A. Berman, L. Addadi, and S. Weiner, *Nature,* **331,** 546 (1988); J. Moradian-Oldak, F. Frolow, L. Addadi, and S. Weiner, *Proc. Roy. Soc. Lond. B.,* **247,** 47 (1992); D. Hanein, B. Geiger, and L. Addadi, *Langmuir,* **9,** 1058 (1993).
7. D. B. DeOliveira and R. A. Laursen, *J. Am. Chem. Soc.,* **119,** 10627 (1997).
8. M. Chalfie and S. Kain, Eds. *Green Fluorescent Protein: Properties, Applications, and Protocols* (Wiley-Liss, Inc., New York, 1998).
9. D. C. Fries, S. T. Rao, and M. Sundaralingam, *Acta Cryst.,* **B27,** 994 (1971); C. A. Beevers and H. N. Hansen, *Acta Cryst.,* **B27,** 1323 (1971).
10. R. A. Visser, *Neth. Milk Dairy J.,* **36,** 167 (1982).
11. M. Kurimoto, P. Subramony, R. W. Gurney, S. Lovell, J. Chmielewski, and B. Kahr, *J. Am. Chem. Soc.,* **121,** 6952 (1999).
12. J. Paquette and R. J. Reeder, *Geology,* **18,** 1244 (1990); P. A. Northrup and R. J. Reeder, *Am. Mineral.,* **79,** 1167 (1994); J. Paquette and R. J. Reeder, *Geochim. Cosmochim. Acta,* **59,** 735 (1995); J. Rakovan and R. J. Reeder, *Geochim. Cosmochim. Acta,* **60,** 4435 (1996); J. Rakovan, D. K. McDaniel, and R. J. Reeder, *Earth Planet. Sci. Lett.,* **146,** 329 (1997).
13. M. Pluta, in *Advanced Light Microscopy: Specialized Methods* (Elsevier, New York, 1989), Vol. 2, Chapter 7.
14. T. D. Dincer, G. M. Parkinson, A. L. Rohl, and M. I. Ogden, *J. Cryst. Growth,* **205,** 368 (1999).
15. L. N. Rashkovich, *KDP-Family Single Crystals,* translated by O. Shlakhova (Adam Hilger, Bristol, 1991).
16. B. Kahr, S.-H. Jang, J. A. Subramony, M. P. Kelley, and L. Bastin, *Adv. Mater.,* **8,** 941 (1996); J. A. Subramony, S.-H. Jang, and B. Kahr, *Ferroelectrics,* **191,** 293 (1997); J. A. Subramony, *PhD Dissertation,* (Purdue University, 1999).
17. B. Kahr, S. Lovell, and J. A. Subramony, *Chirality,* **10,** 66 (1998).
18. For other examples of enantioselective crystal adsorption, see: L. Addadi, Z. Berkovitch-Yellin, I. Weissbuch, M. Lahav, and L. Leiserowitz, *Top. Stereochem.,* **16,** 1 (1986); D. Hanein, B. Geiger, and L. Addadi, *Science,* **263,** 1413 (1994); A. M. Cody and R. D. Cody, *J. Cryst. Growth,* **113,** 508 (1991).
19. Enraf-Nonius KappaCCD Software (Enraf-Nonius, Delft, Netherlands, 1997).
20. G. M. Sheldrick, *SHELX-97-2. Program for the Refinement of Crystal Structures* (University of Göttingen Göttingen, Germany, 1998).
21. J. West, *Zeit. Krist.,* **74,** 306 (1930).
22. W. Kaminsky and A. M. Glazer, *Ferroelectrics,* **183,** 133 (1996).
23. J. Kobayashi and Y. Uesu, *J. Appl. Cryst.,* **16,** 204 (1983).
24. S. Arzt, *Dissertation* (Oxford University, 1995).
25. N. P. Zaitseva, J. J. De Yoreo, M. R. Dehaven, R. L. Vital, K. E. Montgomery, M. Richardson, and L. J. Atherton, *J. Cryst. Growth,* **180,** 255 (1997).
26. *Colour Index* (Society of Dyers and Colourists, London, 1982).
27. N. Zaitseva, L. Carman, I. Smolsky, R. Torres, and M. Yan, *J. Cryst. Growth,* **204,** 512 (1999).
28. W. J. P. van Enckevort, R. J.-V. Rosmalen, and W. H. van der Linden, *J. Cryst. Growth,* **49,** 502 (1980).

29. B. Kahr and R. W. Gurney, submitted for publication.
30. R. W. Gurney, M. Kurimoto, J. A. Subramony, L. D. Bastin, and B. Kahr, *Anisotropic Organic Materials – Approaches to Polar Order, ACS Symposium Series* (in press).
31. R. W. Gurney, C. A. Mitchell, S. Ham, L. D. Bastin, and B. Kahr, *J. Phys. Chem.*, **104,** 878 (2000).
32. O. Lehmann, *Z. Phys. Chem.*, **8,** 543 (1891); *Ann. Phys. Chem.*, **51,** 47 (1894).
33. P. Gaubert, *Bull. Soc. Fr. Min.*, **28,** 286 (1905); *Comptes Rendus*, **142,** 219, 936, (1906); **147,** 632 (1908); **149,** 1004 (1909); **180,** 378 (1925); **204,** 599 (1937).
34. A. Neuhaus, *Angew. Chemie.*, **54,** 527 (1941); *Zeit. Krist.*, **103,** 297 (1941), **105,** 161 (1943); *Zeit. Phys. Chem.*, **A191,** 359 (1943).
35. S. Lovell, *PhD Dissertation* (Purdue University, 2000).
36. C. A. Mitchell, S. Lovell, K. Thomas, P. Savickas, and B. Kahr, *Angew. Chem. Int. Ed. Engl.*, **35,** 1021 (1996).
37. C. J. Drummond, F. Grieser, and T. W. Healy, *J. Chem. Soc., Faraday Trans. I*, **85,** 561 (1989).
38. N Cabrera and D. A. Vermilyea, *Growth and Perfection of Crystals* (Wiley, New York, 1958).
39. A. Mauri, M. Moret, *J. Cryst. Growth,* **208,** 599 (2000).
40. K. Sangwal and S. Veintemillas-Verdaguer, *Cryst. Res. Technol.*, **29,** 639 (1994).
41. H. E. Buckley, *Memoirs Proc. Man. Lit. Phil. Soc.,* **83,** 31 (1938-1939).

Mat. Res. Soc. Symp. Vol. 620 © 2000 Materials Research Society

Growth of the {101} face of KDP crystals in the presence of dye Chicago Sky Blue

Olga A. Gliko, Natalia P. Zaitseva*, Leonid N. Rashkovich
Department of Physics, Moscow State University, Moscow 119899, Russia
*Lawrence Livermore National Laboratory, Livermore, CA 94550, USA

ABSTRACT

In situ interference technique is used for a study of morphology and growth kinetics of a pyramidal face of KDP crystals with dye Chicago Sky Blue. It is shown that the impurity is captured by a crystal when its concentration in the solution exceeds a certain temperature depended value. In the presence of impurity the face loses its stability with increasing of supersaturation. Further increase of supersaturation leads to the restore and subsequent loss of stability. The experimental results are interpreted in the framework of earlier developed model for a crystal growth in the presence of the easily desorbed mobile impurity. The parameters characterizing the adsorption behavior of mobile impurity and its capture have been determined.

INTRODUCTION

It is known that the impurities strongly affect the structural perfection of the crystals grown from the solution. However, the mechanisms of the impurity influence are not studied experimentally well in spite of numerous theoretical works on this subject.

Recently, the number of papers dealing with the growth of KDP crystal in the presence of different impurities was published [1-5]. It was found that the most of cationic impurities affected only the growth of prismatic face, and did not influence the growth of a pyramidal face. For example, the difference in the values of distribution coefficient between solution and growth sectors of these faces for ions of the iron-group exceeds one hundred times. Probably, this fact is due to the positive charge of the face of a pyramid and neutrality of a prismatic face [6]. The anionic impurities have a weak effect on the KDP growth, but some of them slow down the growth of pyramid [4]. However, it was found that the dye Chicago Sky Blue enters the growth sectors of the pyramidal face and intensively stains them [7]. At the same time, the dye is not captured by the prismatic faces. The formula of dye is $C_{34}H_{24}N_6Na_4O_{16}S_4$ - 1-amino-8-naphthol-2,4–disulfonic acid, F.W. = 992.82, and it has 4 negatively charged groups SO_3^- in the solution. These experiments were performed at room temperature with the solutions containing a large amount of the dye by the use of spontaneous crystallization technique. The crystal size did not exceed several millimeters. Some features of incorporation of the dye in the crystal were determined for the large KDP crystals with the linear sizes more than 10 cm grown using the temperature decreasing technique [8]. Nevertheless, it was interesting to study the influence of a given impurity on the KDP crystallization in more detail, because no data on the effect of impurity on the morphology and growth kinetics of the pyramidal face were reported up to now.

TECHNIQUE

The in situ interference technique was used [9]. The crystal plate ($\sim$5x5x2mm^3) cut parallel to the pyramidal face was put into the flow-throw cell, and the solution flow passed over the face. The high enough velocity of solution flow ($\sim$30 cm/s) ensured a kinetic growth regime. The interference image revealing the surface relief was processed by computer. He-Ne laser was used as a light source.

The deionized water with the specific resistance 16 MΩ·cm and salt containing a minimal quantity of nonorganic impurities were used to prepare the solution. The impurity composition of this salt (including 30 elements) is given in reference [3]. In our experiments the dye was introduced into the pure solution at some supersaturation. The supersaturation $\sigma = (c-c_0)/c_0$ (where c and c_0 are the actual and equilibrium concentration in the solution) was created by the reducing of temperature in the interval of 33-37°C.

EXPERIMENT

Growth of large crystals

Crystals were grown in crystallizers with the volumes of about 10 l at the initial temperature ~ 60°C and the impurity concentration $C_i = 1.4$ ppm ($1.4 \cdot 10^{-6}$ mole/mole KDP), which corresponds to ~ 0.33 mg of dye per 1 kg of solution. The solution had an intensive blue color. The growth rate in Z direction was 6-10 mm/day.

It was found that the staining of crystal began only at temperatures below 35°C, when the crystal already has reached a large enough size. It could result from a sharp increase of dye adsorption related to the decrease of temperature. For dye concentrations as low as 0.3 ppm the crystal remained unstained even at 9°C.

The spectrographic analysis of the dissolved part of stained sector of pyramid has shown that the impurity distribution coefficient was close to 1 (the ratio mole of dye/ mole of KDP was the same in the crystal and in the solution). At the same time the growth sectors of prism were not stained.

Figure 1 shows photographs of grown crystals obtained under a different illumination. The dislocation hillock occupying the most part of the pyramidal face and several small hillocks are clearly seen. As known, three vicinal sectors have different slopes. The lower sector at the figure has the steepest slope and the right sector - the shallowest one. The steeper is the sector the smaller is the velocity of growth layers. From figure 1a we see that the staining is weaker as the

Figure 1. *The photographs of crystals grown from the solution with dye. The lines dividing the vicinal sectors of hillocks are clearly seen. The steps on the steepest sector are parallel to the edge with prismatic face. This sector is practically not stained by the dye - a. The macrosteps cover the surface of the bipyramid – b. The size of the edges between the prism and the pyramid faces is about 10 cm.*

steeper is the vicinal sector. Therefore, the impurity capturing is stronger as the higher is the velocity of growth layers. Figure 1b demonstrates macrosteps that are clearly seen over all the surface of dislocation hillock. We point out that such big macrosteps have never been observed in pure crystals. It has also been observed, that impurity incorporation was not uniform, and the stained fragments resembled small spots.

More detailed information was obtained in the process of interference studies.

Surface morphology

Increasing of supersaturation in pure solution (till a certain value $\sigma < 0.08$) does not result in the occurring of macrosteps (figure 2).

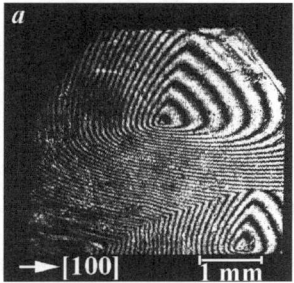 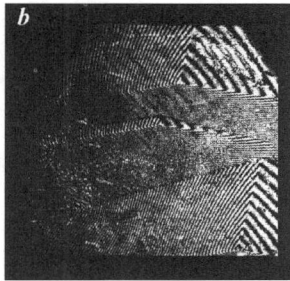

Figure 2. *The surface in pure solution (saturation temperature is 37.6°C); a) $\sigma = 0.02$, b) $\sigma = 0.056$. Developing new hillock and disappearing old one are seen in the top part of the face on the second image.*

The dye introducing into the solution with concentration lower than 1ppm does not effect the surface morphology. At the higher impurity concentration the surface becomes spoiled just after the dye introducing (figure 3). The minimum supersaturation at which the impurity was added was $0.015 - 0.02$ (the corresponding tangential growth rate at the steepest sector $\sim 3 \cdot 10^{-4}$ cm/s). The step regularity is longest maintained near the top of hillocks, but the size of this region decreased gradually with time. Many new small hillocks appeared and then rapidly disappeared in the periphery of the hillock, replacing each other (figure 3b-d). Increase of supersaturation led to enlarging the activity of old hillocks and recovering the stability of surface morphology (figure 3e). However, further increase of σ resulted in the loss of stability once again (figure 3f).

Growth kinetics

Figures 4-6 show the obtained dependencies of the normal (R) and the tangential (v) growth rates of dislocation hillocks as well as their slope (p) on supersaturation in the pure solution and in the solution with the dye.

It is possible to state that the dye does not effect the normal growth rate of a pyramid face (figure 4). From the figure it is clearly seen, that the measured values of R in the pure solution and in the presence of the dye are close. The difference between them can be attributed to the

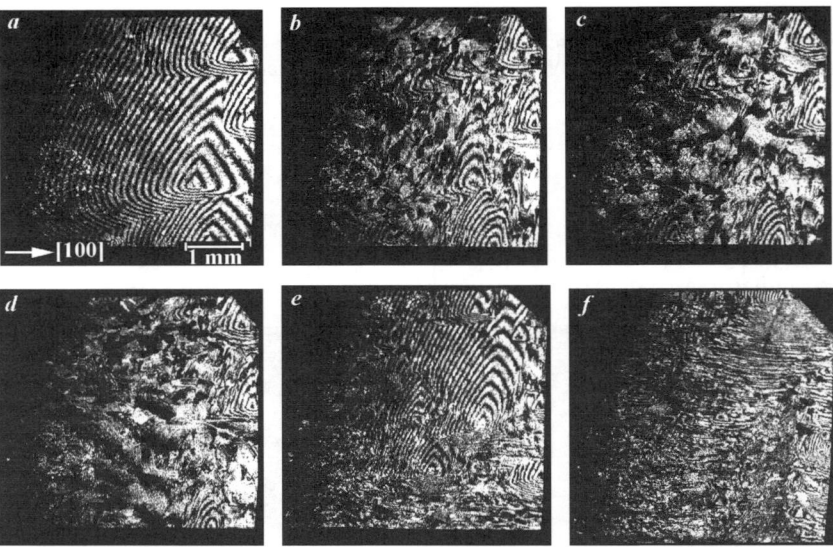

Figure 3. *The influence of dye on the surface morphology.* $C_i = 2ppm$, *saturation temperature is 37.7°C. a) image of the surface in a pure solution, $\sigma = 0.022$, b) image of the same surface 5 min later after adding of dye, c) 10 min later, d) 40 min later; the reducing of the surface of stable regions on old hillocks and appearance of new hillocks are clearly seen; e) $\sigma = 0.034$, the surface is restoring, f) $\sigma = 0.049$, stability loses again.*

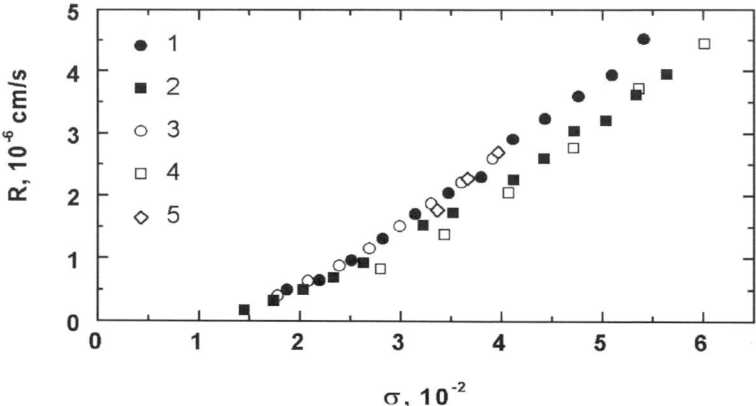

Figure 4. *Influence of supersaturation on the normal growth rate (R) of dislocation hillocks in the solution with different dye concentration. C_i, ppm: 1,2 - 0; 3 – 1; 4 – 1.5; 5 – 2. Saturation temperature, °C: 1,4 - ~33; 2,4,5 - ~37.5.*

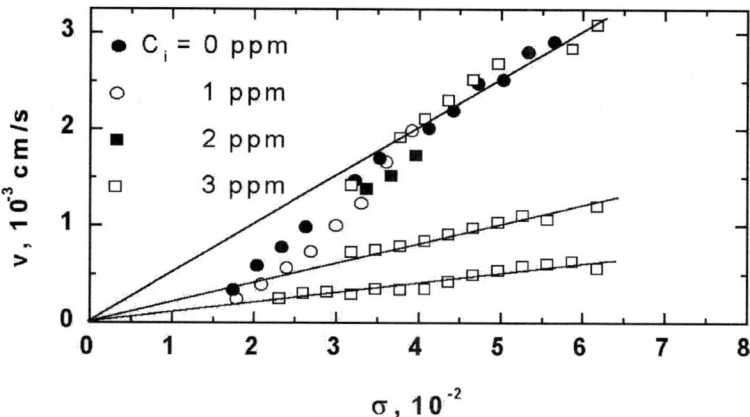

Figure 5. *Influence of supersaturation on the velocity (v) of moving growth layers in three vicinal sectors. Saturation temperature ~37.5⁰C in all experiments.*

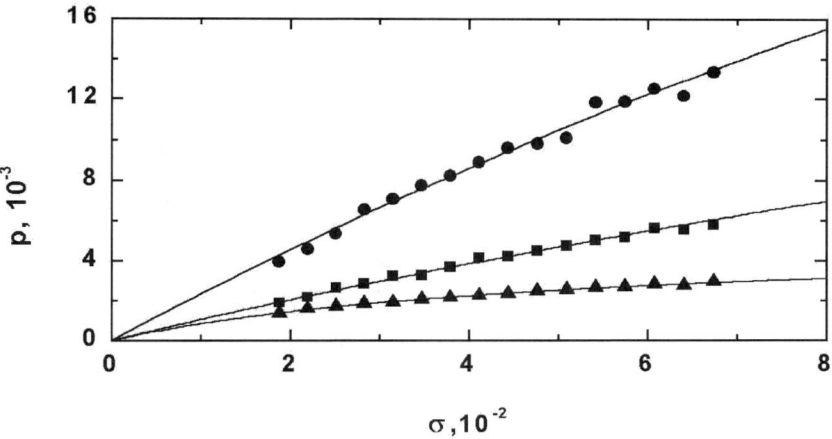

Figure 6. *Influence of supersaturation on slope of vicinal hillocks (p) in pure solution. The curves are drawn to correspond to equation (8).*

difference in the activity of dislocation sources and influence of temperature. In any case, the correlation between R and impurity concentration is absent.

The experimentally derived dependence $v(\sigma)$ has a character unusual for pyramidal faces. The similar behavior of growth rate for pyramid was reported earlier [10]. The dependence $v(\sigma)$

is sublinear for the shallow sector below $\sigma \approx 0.035$, whereas v is nearly proportional to σ for steep sectors in the whole region of supersaturation (figure 5). As it can be seen from figure 4, the dye concentration does not effect noticeably the velocity of the growth layers movement.

The observed ratio between the kinetic coefficients for sectors of different slope and dependencies of slope upon supersaturation $p(\sigma)$ (figure 6) were usual for these faces [10,11].

DISCUSSION

The fact, that the capturing of the dye starts only at a high enough velocity of the step (when the dye does not captured by the steepest vicinal sector, it is captured by the shallow one), gives evidence for a short lifetime of impurity on the surface. Another argument is that the capturing of the impurity occurs only at its high enough concentration in the solution. If the impurity were strongly adsorbed at the surface, the increase of concentration would prevent the capture because the particles of impurity served as obstacles for forward step movement. It seems reasonable to explain the observed phenomena in the framework of earlier developed model describing the movement of a step and its morphological stability in the presence of mobile impurity stoppers [12-14]. Using relationships obtained in these papers we can estimate parameters describing the adsorption behavior of mobile impurity and its capturing.

The main parameters of our model are: the equilibrium degree of covering the sites on the surface (θ), the lifetime of the adsorbed impurity (τ), the surface diffusivity of impurity (D), the capture coefficient of impurity by the moving step K_{st} (the ratio between the impurity concentration at long distance from the step and concentration in the surface layer formed by the step). We suppose that a two-dimensional diffusion of intrinsic growth units does not play significant role because the feed is provided by the incorporation of particle from the solution directly into the step edge [14]. The diffusion length of impurity $l = (D\tau)^{1/2}$ is smaller than half of a distance between the steps. The impurity concentration far from the step is constant and corresponds to θ. The impurity accumulates in front of the moving step where its concentration is determined by the supply of impurity (proportional to the step velocity v) and its removal by the processes of desorption and capturing. The effective supersaturation depends on the impurity concentration in the front of the step, because the mobile impurity can be considered as a two-dimensional gas, whose pressure on the step decelerates it. The more impurity is accumulated, the less is the effective supersaturation [12, 13]. At the higher step velocity the particles of impurities can be considered as immobile relative to the step, forming a classical fence [16]. In this situation the step moves forward by desorption or capturing of stoppers.

Morphology stability

The gradient of impurity concentration normal to the step and related gradient of the effective supersaturation can lead to the enhancement of random perturbation of the step position. This situation occurs if the decrease in the free energy related to the advancement of a local portion of the step in the region of higher σ (smaller impurity concentration) dominates on its increase by means of enlarging of the surface energy of extended step edge. At lower v, when the desorption prevails on the capturing in the process of impurity accumulation, the criterion of loss of stability has a form

$$v > v_n = (D/l)(\alpha\Omega/k_BT/\theta)^{1/3},\qquad(1)$$

where k_B - the Boltzmann constant, T – temperature, Ω - the volume of molecule in the crystal. Instability of the step means that it acquires a cellular structure consisting from the deep concave segments and dividing them convex ones. Maximal depth of concave segments, "pockets", is about of the critical nucleus diameter

$$D_{cr} = 4\alpha\Omega/k_BT\sigma. \tag{2}$$

At the larger depth of the "pocket" the side walls close to each other and the "pocket" collapses. Covering by next steps results in a irregular relief on the whole surface: hillocks and pits replacing each other cover the hillock sectors. On the flat regions occasionally appeared new small hillocks are formed in the emerging points of screw dislocations. Instability of the step develops progressively during its moving away from the top of dislocation hillock, because the development of cellular structure needs a certain time. Irregular relief appears first time on a periphery of hillocks, then it extents into their central parts.

There are two reasons for a spotted character of impurity inclusion. Before the loss of stability when the step has segments-like shape, the mobile stoppers forming the line holding the step, detach the step and diffuse along a bisector of angle between the two adjacent segments. Therefore, the distance between the closely spaced particles of impurity diminishes and as a result the capturing of particles occurs not in a sequential order one by one but the groups of particles are captured at once. When the step loses its stability, the impurity particles are accumulated at the bottom of "pockets" because the characteristic length-scale at which the impurity concentrates is less then the depth of the "pocket".

At higher velocities the accumulation of impurity in front of the step as well as its desorption is limited more and more by capturing. As a consequence, the gradient of the impurity concentration decreases and the step restores its stability. The corresponding criterion is

$$v > v_s = (\theta D/K_{st}^2)(k_BT/\alpha\Omega). \tag{3}$$

The further increasing of velocity in our experiments led again to the loss of stability. It is explained by entering into the field of validity of criterion (1) – in order to increase σ (and v) we decreased temperature and it led to the increase of adsorption.

So, the changes of the surface morphology in the presence of dye at increased supersaturation is described well in the framework of mobile stoppers model. The derivation of criteria (1) and (3) is given in reference [14].

Kinetics

In the case when the effective supersaturation is determined by capturing of impurity and $K_{st} \ll 1$, as it is shown in references [12, 13], the dependencies $v(\sigma)$ shift into the field of higher supersaturation compared with dependencies for the pure solution. The value of this shift is evaluated in accordance with

$$\sigma_d = \theta/K_{st}. \tag{4}$$

This relationship is valid at the condition

$$v > v_d = (D/\tau K_{st})^{1/2}. \tag{5}$$

For estimation of parameters σ_d and D/θ we use the following data obtained from experiment (see figure 3e) at $C_i = 2$ppm and $t \approx 37\,°C$: $v_n \approx 4\cdot10^{-4}$ cm/s, $v_s \approx 1.2\cdot10^{-3}$ cm/s (corresponding value of σ is 0.034). The lifetime of impurity can be evaluated as a time necessary for step to pass around the particle and capture it: $\tau \approx D_{cr}/v$.

D_{cr} is estimated from the equation (2). At $\alpha = 20$ erg/cm^2 we calculate $\alpha\Omega/k_BT = 4.5\cdot10^{-8}$ cm. At minimal supersaturation $\sigma = 0.0175$ and $v = 3\cdot10^{-4}$cm/s we obtain $\tau < 3.4\cdot10^{-2}$s. At such value of σ the exposition time of the terrace between the steps $h/R \approx 0.3$ s (here $h = 5.1\cdot10^{-8}$ cm is the step height, $R \approx 2\cdot10^{-7}$ cm/s) is 10 times greater then the lifetime of adsorbed impurity.

When the values v_s, v_n and τ are known, we can estimate the parameters spoken above (σ_d and D/θ). From (1) and (3) taking into account that $l^2 = D\tau$, we obtain:

$$\sigma_d = \theta/K_{st} = (\Omega\alpha/k_BT)(1/v_n\tau)(v_s/v_n)^{1/2} \approx 5\cdot10^{-3}, \tag{6}$$

$$D/\theta = v_n^3\tau^2/(\Omega\alpha/k_BT) \approx 1.5\cdot10^{-6} \text{ cm}^2/\text{s}. \tag{7}$$

Low value of σ_d means that it is practically impossible to observe a difference between the behavior of $v(\sigma)$ for pure and doped solution. Assuming $\theta \approx 10^{-6}$ we get $K_{st} = 2\cdot10^{-4}$, $D = 1.5\cdot10^{-12}$ cm^2/s and $l = 2.3\cdot10^{-7}$ cm. Note that the distance between the steps h/p in the shallow sector varies around from $5\cdot10^{-3}$ to $1.5\cdot10^{-3}$ cm, i. e. is sufficiently larger then the diffusion length of the impurity. The value l is also less then D_{cr}. Beside this, $v_d = 1.5\cdot10^{-6}$cm and it means that the condition (5) is satisfied. All given estimations do support our ideas on the effect of mobile impurity on face morphology and step kinetics.

Finally, consider the data illustrated by figure 6. The all three experimental dependencies $p(\sigma)$ are in a good fitting with the equation

$$p = \sigma/(A + B\sigma), \tag{8}$$

where A and B are constants, calculated by the least-square technique. The equation (8) corresponds to the dependence $p(\sigma)$ theoretically predicted for the model taking into account the effect of complex dislocation source, presence of dislocation core and linear dependence of v upon σ for all sectors [11]. In this situation the relationship between the slope of different sectors must be not dependent on supersaturation. However, as it is seen from figure 6, it is hold only for steep sectors. The case when the dependence $v(\sigma)$ is nonlinear only for one sector has not been analyzed till now. Such a situation for an isotropic spiral leads to the appearance of minimum on the curve $p(\sigma)$ [15], therefore it seems to be strange that the experimental data for the shallow sector satisfies (8). This problem needs a special consideration.

CONCLUSIONS

On the example of dye Chicago Sky Blue adsorbed by a pyramid face of KDP crystal we have demonstrated that the impurity easily desorbed and not very mobile on the surface does not practically effect the kinetics of steps. At the same time, the impurity accumulating in front of the step results in creating of gradient of effective supersaturation and leads to the loss of morphological stability of steps echelon in certain range of step velocities. As the velocity increases further, the step loses its stability once again due to the increase of the surface concentration of impurity related to decrease of temperature. The obtained results provide the experimental confirmation to the earlier developed theoretical model.

ACKNOWLEGMENTS

The authors acknowledge B. Kahr who initiated this study and V.V. Voronkov for fruitful discussion. This work was supported in part by the Russian Foundation for Basic Research (grants 99-02-16319 and 00-02-16701) and NATO (grant PST.CLG 975240).

REFERENCES

1. A.A. Chernov, L.N. Rashkovich, *J. Crystal Growth*, **84**, 389 (1987).
2. L.N. Rashkovich, B.Yu. Shekunov, in: Growth of Crystals, Vol.18, Eds. E.I. Givargizov and S.A. Grinberg (Plenum, New York, 1990) p.124.
3. L.N. Rashkovich, N.V. Kronsky, *J. Crystal Growth*, **182**, 434 (1997).
4. Y.-J. Fu, Z.-S. Gao, J.-M. Liu, Y.-P. Li, H. Zeng, M.-H. Jiang, *J. Crystal Growth*, **198/199**, 682 (1999).
5. T.A. Land, T.L. Martin, S. Potapenko, G.T. Palmor, J.J. DeYoreo, *Nature*, **399**, 442 (1999).
6. S.A. de Vries, P. Goedtkindt, W.J. Huisman, M.J. Zwanenburg, R. Feidenhans'l, S.L. Bennett, D.-M. Smilgies, A.Stierle, J.J. De Yoreo, W.J.P. van Enckevort, P. Bennema, E. Vlieg, *J. Crystal Growth*, **205**, 202 (1999).
7. J.A.Subramony, S.-H. Jang and B.Kahr, *Ferroelectrics*, **191**, 293 (1997).
8. N. Zaitseva, L. Carman, I. Smolsky, R. Torres, M. Yan, *J. Crystal Growth*, **204**, 512 (1999).
9. L.N. Rashkovich, A.A. Mkrtchan, A.A. Chernov, *Sov.Phys.-Cryst.*, **30**, 219 (1985).
10. L.N. Rashkovich, G.T. Moldazhanova, in: Growth of Crystals, Vol.20, Eds. E.I. Givargizov and A.M. Melnikova (Consultants Bureau, New York, 1996) p.69.
11. J.J. De Yoreo, T.A. Land, L.N. Rashkovich, T.A. Onischenko, J.D. Lee, O.V. Monovskii, N.P. Zaitseva, *J. Crystal Growth*, **182**, 442 (1997).
12. V.V. Voronkov, L.N. Rashkovich, *Sov. Phys.-Cryst.*, **37**, 559 (1992).
13. V.V. Voronkov and L.N. Rashkovich, *J. Crystal Growth*, **144**, 107 (1994).
14. V.V. Voronkov, in: Growth of Crystals, Vol.20, Eds. E.I. Givargizov and A.M. Melnikova (Consultants Bureau, New York, 1996) p.59.
15. L.N. Rashkovich, KDP – family Single Crystals (Adam Hilger, Bristol, Philadelphia and New-York, 1991).
16. N. Cabrera, D.A. Vermilyea, in: Growth and Perfections of Crystals, Eds. R.H. Doremus, B.W. Roberts, D. Turnball (Wiley, New York, 1958) p. 393.

AUTHOR INDEX

SUBJECT INDEX